MATHEMATICS
Outline and Review Problems for

Other Books by the Author

Applications of Electronics (with Milton S. Kiver)
Basic Television

MATHEMATICS

Outline and Review Problems for

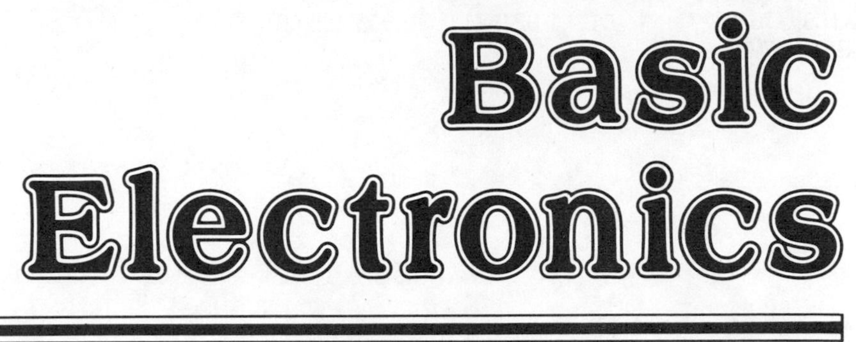

fourth edition

BERNARD GROB

Instructor, Technical Career Institutes, Inc.
(formerly RCA Institutes, Inc.)

Gregg Division
McGraw-Hill Book Company

New York
St. Louis
Dallas
San Francisco
Auckland
Bogotá
Düsseldorf
Johannesburg
London
Madrid
Mexico
Montreal
New Delhi
Panama
Paris
São Paulo
Singapore
Sydney
Tokyo
Toronto

Library of Congress Cataloging in Publication Data
Grob, Bernard.
Mathematics: outline and review problems for basic electronics, fourth ed.

1. Electric engineering—Mathematics. 2. Electronics—Mathematics. I. Title.
TK153.G76 1977 512'.1'0246213 77-22252
ISBN 0-07-024924-5

MATHEMATICS Outline and Review Problems for Basic Electronics. Fourth Edition

1 2 3 4 5 6 7 8 9 0 E B E B 7 8 3 2 1 0 9 8 7

The editors for this book were Gordon Rockmaker and Alice V. Manning, the designers were Marsha Cohen and Tracy A. Glasner, the art supervisor was George T. Resch, and the production supervisor was Iris A. Levy. It was set in Souvenir Light by Progressive Typographers. Printed and bound by Edwards Brothers Incorporated.

Contents

Preface

The purpose of this book is to help students who have trouble with the numerical calculations usually encountered in electrical and electronics work. On the basis of years of teaching experience, it seems that the difficulties can be traced to a combination of three things. Most students have a basic knowledge of the operations of arithmetic, but very often the fine points have been forgotten or never learned. These details of arithmetic are highlighted in this book with illustrative examples and practice problems designed to develop a close working familiarity with the subject.

Another difficulty many students have is in working with literal numbers, that is, the symbol letters used in equations and formulas. Again, this book provides a very concise outline of the operations and manipulations of literal numbers and their use in equations.

Finally, many students studying electricity and electronics have had very little exposure to trigonometry. However, the analysis of ac circuits depends on the ability to solve problems involving triangles, and the best way to solve such problems is through trigonometry. This includes the basic sine, cosine, and tangent functions together with their application to simple circuit problems.

Other key topics include working with decimal numbers, adding and subtracting fractions, using reciprocals, arithmetic operations with negative numbers, finding powers and roots of numbers, and using powers of 10 and the common metric prefixes found in electricity and electronics.

In each case the subject matter is divided into many small steps, similar to the style of programmed instruction, with each step building on the one immediately preceding it. Each step has one or more illustrative examples and numerous practice problems. The answers to the problems are given at the back of the book.

Although specifically written to complement the author's text *Basic Electronics,* this outline is general enough to satisfy the needs of all students of electronics.

Bernard Grob

MATHEMATICS
Outline and Review Problems for

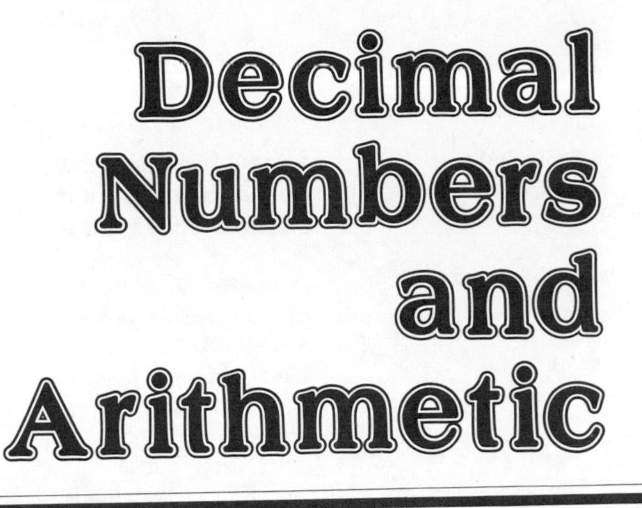

The system of counting and calculating that we learned in school is based on tens, since 10 digits are used. They are 0, 1, 2, 3, 4, 5, 6, 7, 8, and 9. We say such a number system uses the *base 10.* Methods of using these numbers are explained in the following topics:

1-1 THE ORDER OF PLACES

To count values larger than 9, each position to the left increases in value in multiples of 10. The first position is for *units* up to 9. The next place to the left is for *tens,* as in 10, 20, 30, 40, 50, 60, 70, 80, and 90. For example, the digit 2 in this place indicates 20.

The next place to the left after the tens position is for *hundreds.* The digit 3 in this place indicates 300. This order continues with thousands and higher multiples of 10 for each succeeding place to the left.

Example. Indicate the value of each place in the number 4321.

Answer. The 1 unit in the right-hand place is the units digit.

The next digit to the left is the tens digit, and in this case it represents 20.

The next digit to the left is the hundreds digit, and in this case it represents 300.

Finally, the left-hand place is the thousands digit, which in this case represents 4000.

In summary,

4321 = 4000 + 300 + 20 + 1

or

```
 4000
  300
   20
    1
 ----
 4321
```

Problems 1-A

How many places are used in the following numbers?

1.	32	4.	20
2.	302	5.	16,666
3.	9999	6.	5,423,628

Note that a count up to nine can be used in each place, as in 9999. Note also that a zero is used to fill in a place to show no count there, as in 302 and 20. In 302, there is no count for the tens.

1-2 THE DECIMAL POINT

In the preceding section we began counting the position of each digit from right to left. Although not shown, it was understood that a *decimal point* was placed just to the right of the units digit. The 4321 is actually

4321.

The value has not been changed in any way by indicating the decimal point. The decimal point simply marks where the places begin.

Places to the left of the decimal point are for numbers larger than 1. More places to the left of the decimal point mean a larger number. For instance, 2,000,000 is more than 20.

Example. Write the following as a number, and locate the decimal point:

1 unit, 2 tens, 3 hundreds, and 4 thousands

Answer.

```
1 unit      =    1
2 tens      =   20
3 hundreds  =  300
4 thousands = 4000
              ----
     Number = 4321.
```

We can add a zero after the decimal point, as in 4321.0, in order to show there is no further count here.

Problems 1-B

Write each of the following as a number, and show the decimal point:

1. 2 tens plus 5 units
2. 2 hundreds plus 2 tens and 5 units
3. 2 hundreds plus 5 units
4. 3 thousands plus 7 units

1-3 DECIMAL FRACTIONS

Places to the right of the decimal point are used for fractional values of less than 1, in multiples of tenths. One tenth is 0.1 or $^1/_{10}$. The first place to the right of the decimal point is for tenths. For instance, the digit 3 in this place, or 0.3, is equal to 3 tenths, or $^3/_{10}$. Note that for numbers less than 1 the practice in many technical fields, such as electronics, is to write a zero to the left of the decimal point to emphasize there is no count in the units place.

The number 0.3 is a *decimal fraction,* while $^3/_{10}$ is a *common fraction.* More places to the right mean a smaller decimal fraction. For instance, 0.05, or $^5/_{100}$, is less than 0.5, or $^5/_{10}$. Places to the right of the decimal point are always fractional parts of one unit, to represent a number less than 1.

The second place to the right of the decimal point is for hundredths, the third place is for thousandths, and so on.

Example. What decimal places are represented by the number 0.234?

Answer. The first place to the right of the decimal point indicates tenths: $^2/_{10}$, or 0.2.

The next place to the right indicates hundredths: $^3/_{100}$, or 0.03.

The final place to the right indicates thousandths: $^4/_{1000}$, or 0.004.

The complete decimal fraction, then, is

$$0.2 + 0.03 + 0.004 = 0.234$$

Problems 1-C

Write each value as a decimal number.

1. 6 tenths
2. 6 tenths plus 2 hundredths
3. 7 thousandths
4. 8 units plus 3 tenths
5. 2 tenths plus 5 thousandths
6. 7 millionths

1-4 ADDITION

The first step in adding decimal numbers is to line up the decimal points in column fashion. Otherwise, you could be adding together numbers in different decimal places, which would result in the wrong count.

Example. Add 465, 32, and 26,400.

Answer. Remember: the decimal point is assumed to be at the right of the units digit.

```
   456
    32
26,400
------
26,888
```

Example. Add 123.4000, 0.1265, and 6.0001.

Answer. Line up the decimal points and add:

```
123.4000
  0.1265
  6.0001
--------
129.5266
```

In all cases, the decimal points must be lined up so that the values for the same place are in the same vertical column. When one column adds to more than 10, keep the units but carry the multiple of 10 to the next column.

Example. Add 14 + 18.

Answer.

```
14
18
--
32
```

Problems 1-D

Do the following additions:

1. 200 + 30 =
2. 232 + 1000 =
3. 847 + 42 =
4. 682 + 82 =
5. 432.5 + 3.2 =
6. 432.0 + 0.5 =
7. 764.8 + 31.6 =
8. 634.00 + 0.34 =

1-5 SUBTRACTION

As in addition, before doing a problem in subtraction make sure the decimal points in the two numbers are lined up.

Example. Subtract 73 from 125. The decimal point is assumed to be after the units digit.

Answer.

```
 125
- 73
----
  52
```

Example. Subtract 0.1520 from 9.0763.

Answer.

```
  9.0763
- 0.1520
  8.9243
```

Note that in the first column after the decimal point, the 0 at the top was increased to 10 tenths by borrowing 1 from the units column to make it 8 instead of 9.

Problems 1-E

Do the following subtractions:

1. 6.4 – 4.2 =
2. 297 – 293 =
3. 1625.02 – 0.19 =
4. 85 – 61 =
5. 82 – 65 =
6. 1575 – 1009 =
7. 0.0126 – 0.0050 =
8. 1.10 – 1.02 =
9. 200 – 137 =
10. 1,265,422 – 1,006,978 =

1-6 MULTIPLICATION

To set up a problem in multiplication it is not necessary to line up the decimal points of the two numbers. For convenience, however, the right-hand digits of the two numbers are lined up. Each digit of one number is multiplied by each digit of the other number, and the resulting products are added.

Example. Multiply 13.4 by 21.

Answer. The problem can be arranged in either of two ways:

```
  13.4     or     2.1
×  2.1         × 13.4
```

Both arrangements will lead to the same answer, but using the left-hand arrangement requires less calculation. When numbers are multiplied, the answer is called the *product*.

To carry through the multiplication in this example, multiply each digit of the first number in order by the 1 in the second number. Line up the right-hand digit of the product with the 1. Ignore the decimal points at this stage.

```
  13.4
×  2.1
  13 4
```

Next multiply each digit in the top number by the 2 in the second number. Again, ignore the decimal point. Line up the second product with the 2 in the second number, because now this is the digit you are using for multiplication.

```
  13.4
×  2.1
  13 4
 268
```

Finally, add the two products:

```
  13.4
×  2.1
  13 4
 268
 281 4
```

To locate the position of the decimal point in the answer, add the number of digits to the right of the decimal point in each of the two numbers that were multiplied. The first number has one digit to the right of the decimal; the second number also has one digit to the right. Therefore, the final product will have two digits to the right of the decimal point. The answer is

28.14

It should be noted that multiplication can be indicated in three ways as follows:

$6 \times 4 =$ $\quad 6 \cdot 4 =$ $\quad (6)(4) =$

For all three of these examples, the product is 24.

Problems 1-F

Do the following multiplications:

1.	$3.2 \times 3 =$	5.	$400 \times 0.2 =$
2.	$32 \times 0.3 =$	6.	$0.2 \times 400 =$
3.	$42 \times 0.2 =$	7.	$11.1 \times 9 =$
4.	$12.3 \times 2.2 =$	8.	$420 \times 2.2 =$

1-7 DIVISION

The form of a division problem depends on the difficulty of the operation. If we needed to divide 27 by 9, the problem could be set up as

$$^{27}/_{9} = 3 \quad \text{or} \quad \frac{27}{9} = 3 \quad \text{or} \quad 27 \div 9 = 3$$

We know the answer is 3 because we have memorized the fact that $3 \times 9 = 27$.

When a division problem is more complicated, a different form of the operation is used.

Example. Divide 168 by 12. This type of problem can be written

$12\overline{)168}$

Answer. The answer is obtained by dividing 16 by 12 and carrying the remainder of 4 to the next place, as follows:

```
    14
12)168
   12
    48
    48
     0
```

Continue the divisions until the remainder is zero in this example. The answer is exactly 14 because $12 \times 14 = 168$ exactly. The 168 is the *dividend,* 12 is the *divisor,* and the answer of 14 is the *quotient.* The dividend is divided by the divisor.

It should be noted that although it does not matter which of the two numbers is the multiplier in the operation of multiplication, the dividend must be divided by the divisor in the operation of division. For instance, $6 \div 3 = 2$, but $3 \div 6 = {}^{3}/_{6}$, ${}^{1}/_{2}$, or 0.5.

When there is a decimal point to consider in division, the decimal point in the quotient is lined up with the decimal point in the dividend.

Example. Divide 38.4 by 12.

Answer. This is solved as follows:

```
    3.2
12)38.4
   36
    24
    24
     0
```

There should not be any decimal fraction in the divisor. If necessary, move the decimal point to the right in the divisor to eliminate any decimal fraction and move the decimal point the same number of places in the dividend. The value of the quotient is not changed by this procedure, because the divisor and dividend are both multiplied by the same number.

Example. Divide 3.84 by 1.2.

Answer. This is solved as follows:

$1.2\overline{)3.84}$ becomes $12\overline{)38.4}$

The answer here is 3.2, which is the same quotient as in the previous example.

In some cases of division, the answer will not be an exact whole number. Then zeros can be added to the right of the decimal point in the dividend, to as many places as necessary.

Example. Divide 228 by 16.

Answer. This is solved as follows:

```
   14.25
16)228.00
   16
    68
    64
     40
     32
      80
      80
       0
```

In cases where the dividend is not exactly divisible by the divisor, a remainder of zero cannot be obtained. Then you have these three possibilities:

1. When the remainder is less than one-half the divisor, consider the remainder as zero and drop it.
2. When the remainder is more than one-half the divisor, make the last digit in the quotient one more.
3. When the remainder is exactly one-half the divisor, make the next and last digit in the quotient a 5. This corresponds to $^1/_2$, or 0.5, for the last decimal place.

Problems 1-G

Do the following divisions:

1. $384 \div 12 =$
2. $38.4 \div 12 =$
3. $3.84 \div 12 =$
4. $46 \div 0.2 =$
5. $15 \div 5 =$
6. $15 \div 0.5 =$
7. $144 \div 6 =$
8. $0.125 \div 0.25 =$
9. $30 \div 15 =$
10. $15 \div 30 =$

1-8 MOVING THE DECIMAL POINT

In our number system based on ten, we can multiply by 10 or 100 or 1000 merely by moving the decimal point one, two, or three places to the right.

Example. Multiply 4.87 by 1000.

Answer. We can multiply this out by the usual method:

```
   1000
 × 4.87
   7000
  8000
4000
4870.00
```

Notice that the original digits 4, 8, and 7 were unchanged. The decimal point just moved three places to the right.

The rule for multiplying any number by a power of 10 (that is, 10, 100, 1000, 10,000, etc.) is to move the decimal point to the right as many places as there are zeros in the multiplier. When multiplying by 10, move the decimal point one place to the right; when multiplying by 100, move the decimal point two places to the right; when multiplying by 1000, move the decimal point three places to the right; and so on.

This rule also applies to numbers less than 1. The decimal point moves to the right for multiplication, whether the number is greater than 1 or less than 1.

Example. Multiply 0.0042 by 100.

Answer. Since 100 has two zeros, the decimal point is moved two places to the right.

$0.0042 \times 100 = 0.42$

It is not necessary to retain all the zeros to the left of the decimal point. A single zero is enough to indicate that the number is less than 1.

Problems 1-H

Do the following multiplications:

1. $2 \times 100 =$
2. $5.8 \times 100 =$
3. $3 \times 1000 =$
4. $5.432 \times 1000 =$
5. $0.0057 \times 1000 =$
6. $0.000036 \times 1000 =$
7. $4.2 \times 10 =$
8. $0.42 \times 10 =$

In the case of division by powers of 10, the decimal point is moved to the left as many places as there are zeros in the divisor. This procedure for division is the opposite of that for multiplication. In summary, the rules are as follows:

For multiplication, move the point to the right. →
For division, move the point to the left. ←

Example. Divide 642 by 100.

Answer. Since 100 has two zeros, the decimal point in the original number is moved two places to the left.

$642 \div 100 = 6.42$

This method of division also applies to numbers less than 1.

Example. Divide 0.42 by 100.

Answer. Move the decimal point two places to the left.

$0.42 \div 100 = 0.0042$

Problems 1-I

Do the following divisions:

1. $200 \div 100 =$
2. $580 \div 100 =$
3. $3000 \div 1000 =$
4. $5432 \div 1000 =$
5. $0.57 \div 100 =$
6. $0.036 \div 1000 =$
7. $42 \div 10 =$
8. $4.2 \div 10 =$

1-9 SHORTCUTS WITH MULTIPLES OF 10

In some other cases, too, moving the decimal point is a shortcut for multiplication or division.

Example. Multiply 44 by 50.

Answer. Since $50 = {}^{100}/_{2}$, we can use the shortcut of moving the decimal point by looking at the problem as

$$\frac{44 \times 100}{2}$$

We know from the decimal-moving process that $44 \times 100 = 4400$. We then need only divide this product by 2, obtaining

$$\frac{4400}{2} = 2200$$

As a result, for the final answer,

$$44 \times 50 = 2200$$

This kind of operation is also possible with division.

Example. Divide 44 by 50.

Answer. Since $50 = {}^{100}/_{2}$, we can use the shortcut of moving the decimal point by looking at the problem as

$$44 \div \frac{100}{2}$$

We know from the decimal-moving process that $44 \div 100 = 0.44$. Remember: the decimal point is moved two places to the left for division by 100 here, which is the opposite of two places to the right for multiplication by 100.

Now we have $0.44 \div 1/2$. The $1/2$ comes from the 2 in the denominator of the divisor $100/2$. To divide by a fraction, it is inverted and multiplied, as follows:

$$0.44 \div \frac{1}{2} = 0.44 \times \frac{2}{1} =$$

$$= 0.88$$

As a result, for the final answer,

$$44 \div 50 = 0.88$$

Problems 1-J

Do the following multiplications and divisions:

1. 44×500
2. 24×50
3. 12×25
4. $44 \div 500$
5. $24 \div 50$
6. $12 \div 25$

1-10 SQUARES AND SQUARE ROOTS

The square of a number is that number multiplied by itself. When we say "square 5" or "5 squared," it just means 5×5, for the answer of 25. This is generally written as $5^2 = 5 \times 5 = 25$.

The 5 here is the *base* number. The 2 written above is the *exponent* or *power* of the base number. For the exponent of 2, the base number is squared or raised to the second power. Any number can be the base.

Examples. $3^2 = 3 \times 3 = 9$
$10^2 = 10 \times 10 = 100$

The square root of a number is the number that when multiplied by itself equals the original number. This *radical sign* $\sqrt{\ }$ shows that we want to find the square root.

Example. $\sqrt{25} = 5$

Answer. The proof is that $5 \times 5 = 25$. Therefore, 5 is the square root of 25.

The purpose of using exponents is to have a shortcut method for continued multiplication of the same number. Actually, any number can be the base, raised to any power. These operations are described for cubes as the third power, and for all powers and roots in general, in Chap. 4. Exponents are especially useful as powers of 10 to keep track of the decimal point for very large or small numbers, as described in detail in Chap. 5.

Problems 1-K

Find the square or square root.

1. 3^2
2. 4^2
3. 5^2
4. 7^2
5. $\sqrt{9}$
6. $\sqrt{16}$
7. $\sqrt{25}$
8. $\sqrt{49}$

1-11 AVERAGE VALUE

When a quantity has different values at different times, it is useful to consider one value as typical in order to have a specific measure. The value probably used most often is the *arithmetical average*. This value equals the sum of all the values divided by the number of values.

Example. The test scores for a student are 70, 80, and 90. Find the student's average score.

Answer. To find the average, add the test scores and divide by the number of scores:

$$\begin{array}{r} 70 \\ 80 \\ \underline{90} \\ 240 \end{array} \qquad \begin{array}{r} 80 \\ 3\overline{)240} \end{array}$$

The average score is 80. This happens to be the middle, or *median,* score also. The arithmetic average is also called the *arithmetic mean.*

Example. Find the average of 60, 70, 80, and 80.

Answer. For the addition,

$$60 + 70 + 80 + 80 = 290$$

For the division,

$$4\overline{)290.0} = 72.5$$

The average of 60, 70, 80, and 80 is therefore 72.5.

Average calculations can misrepresent the actual physical conditions that are present in a problem, although they may be mathematically correct. If one number is very much different from the others in a group, the average value may be unduly weighted toward this value.

Example. The following voltage readings were made in a circuit. Find the average voltage.

110, 112, 115, 115, 27

Answer. Following the usual procedure, we find the average:

$$\begin{array}{r} 110 \\ 112 \\ 115 \\ 115 \\ \underline{27} \\ 479 \end{array}$$

$$5\overline{)479.0} = 95.8$$

Although our calculations found an average value of 95.8 volts, this is not really the condition of the circuit. It is obvious that the 27-volt reading has brought the average down below a *reasonable* value. Therefore, in finding an arithmetic average that must relate to some real physical condition, such as the readings in an experiment, it may be necessary to discard any value that is vastly different from the others in the group. It is a good idea to check on what caused the unreasonable value.

Problems 1-L

Find the arithmetical average for the following groups of values:

1. 13, 8, and 9
2. 70, 80, 90, and 95
3. 0.5, 0.7, 0.4, and 0.8
4. 0 V, 33 V, 50 V, 78 V, 88 V, 97 V, and 100 V (*Note:* V is for volts.)
5. 100 Ω, 75 Ω, 83 Ω, 1000 Ω (*Note:* the Ω is for ohms of resistance.)

1-12 ROOT-MEAN-SQUARE (RMS) VALUE

In some applications, the squares of the individual values are important for the average. An example is electrical power, with the formula I^2R. This means the power depends on the square of the current intensity I. Alternating current and voltage are usually specified by their *root-mean-square,* or *rms,* value.

An rms value is derived by first taking the square of each individual value, adding the squares, and dividing the sum by the number of values. This results in a mean of the squares. The square root of this mean is the rms value.

Example. Find the rms value for 2, 3, and 4.

Answer. The squares are 4, 9, and 16. Their sum is 29. The mean square is

$$29/3 = 9.67$$

The square root of 9.67 is 3.11. As a result, the answer for the rms value of 2, 3, and 4 is 3.11.

Problems 1-M

Find the rms value for the following groups of values:

1. 3, 4, 5, 6, and 7
2. 5, 7, 9, and 10
3. 2, 3, and 3
4. 20, 30, and 40

1-13 SIGNS FOR GROUPING NUMBERS

In order to show the sequence that should be used in combining numbers, the following mathematical symbols are used:

Parentheses ( )
Brackets [ ]
Braces { }

Parentheses surround the lowest grouping. The bracket is the next higher symbol and can surround any groupings of parentheses. The brace is the highest symbol. A pair of braces can surround groupings of brackets.

Example. Do the calculations indicated by the grouping symbols.

$\{4[(2 + 3)^2 + 6] - 5\} =$

Answer: The first calculation is performed in the parentheses:

$2 + 3 = 5$

Then $(5)^2 = 25$

Next the calculation involving the bracket is performed:

$4[25 + 6] = 4[31] = 124$

Finally, the calculation in the braces is performed:

$124 - 5 = 119$

When a group has multiplications and divisions or powers and roots, these operations must be completed before the groups can be added or subtracted.

Examples. $4 + (4 \times 2) = 4 + 8 = 12$

Also: $35 - (2 + 3)^2 = 35 - 25 = 10$

For division, the fraction bar can be considered a sign of grouping. You cannot divide until all the additions or subtractions are combined in either the numerator or denominator or both.

Example. $2 + \frac{4 + 5}{3} = 2 + \frac{9}{3} = 2 + 3 = 5$

Problems 1-N

Do the following combined operations:

1. $6 + (2 \times 3) =$
2. $(14 \times 2) - 20 =$
3. $\frac{8 + 12}{5} + 3 =$
4. $5\left[2 + \frac{1}{2}(3 + 1) - 4\right] =$
5. $5(3 + 6) =$
6. $4 + (8 - 3)^2 =$
7. $3(2 + 2) =$
8. $\left(\frac{4 + 8}{6} + 3\right) \times 6 =$

1-14 EVALUATION OF FORMULAS

To solve for the unknown factor in a formula, just substitute the known values for the letters in the formula.

Example. If the sides of a rectangle are 4 inches and 3 inches, solve the formula for the area of a rectangle $A = a \times b$, where a and b are the lengths of the sides of the rectangle and A is the area.

Answer. $A = a \times b$
$A = 4 \times 3$
$A = 12$ square inches

Problems 1-O

Find V in volts with the Ohm's law formula $V = IR$ for the following values of I in amperes and R in ohms. The IR means $I \times R$.

1. $I = 2, R = 5$
2. $I = 5, R = 2$
3. $I = 0.003, R = 3000$
4. $I = 0.003, R = 100$
5. $I = 10, R = 47$
6. $I = 0.000\,002, R = 10{,}000$
7. $I = 3, R = 4000$
8. $I = \sqrt{2}, R = 10$

Problems 1-P

Find P in watts from the power formula $P = I^2R$ for the following values of I in amperes and R in ohms. The I^2R means $I^2 \times R$.

1. $I = 2, R = 5$
2. $I = 5, R = 2$
3. $I = 0.1, R = 100$
4. $I = 10, R = 10$
5. $I = \sqrt{2}, R = 10$
6. $I = 9, R = 3$

1-15 ROUNDING OFF A NUMBER

Sometimes we want to "round off" numbers because we do not need the accuracy indicated by all the digits. This is especially true with electronic calculators. There is no need to work with a value of 36.241 102 31 volts, as an example, when the closest we can measure it is probably 36.24 volts.

Example. Round off 469 to the nearest tens place.

Answer. The answer here is 470, but let us see why. By just dropping the 9 in the units place, the number 460 is nine less than the actual value. On the other hand, 9 in the units place is very close to 10 for the next digit in the tens place. Therefore, raising the 6 to 7 in the tens place for the value of 470 makes the number only one more than the original. Thus 469 accurate to the nearest tens place is 470.

Example. Round off 461 to the nearest tens place.

Answer. The answer here is 460. The 1 in the units place can be changed to zero, without raising the digit in the tens place, as 460 is closer to the original number than 470.

The general rules are as follows:

1. When the digit dropped is 6 or more, raise the last digit by 1.
2. When the digit dropped is 4 or less, keep the last digit the same.

It is important to remember that the digit dropped must be replaced by a zero. This procedure keeps the same number of decimal places in the number.

Examples. Rounded off to the nearest hundreds place,

4611 = 4600
4644 = 4600
4664 = 4700
4691 = 4700

All four examples are rounded off to two significant figures, which is the number of digits other than zero before the decimal place in the rounded number.

When the digit to be dropped is 5, there are two possibilities:

1. The 5 is followed by digits more than zero. This tips the value to more than one-half the decimal place. Then raise the digit before the 5 by one.

2. The 5 is followed by one or more zeros. This value is exactly one-half the decimal place. Then you can raise the previous digit only if it becomes an even number.

Examples. Rounded off to the nearest hundreds place,

4651 = 4700 4750 = 4800
4659 = 4700 4650 = 4600

When rounding off decimal fractions less than 1, the procedure is the same. However, the zeros following the rounded digits can be eliminated, because they do not affect the value.

Examples. Rounded off to the nearest hundredths place,

0.4611 = 0.4600 = 0.46
0.4664 = 0.4700 = 0.47
0.4651 = 0.4700 = 0.47

Problems 1-Q

Round off the following to the number of significant digits shown in brackets:

1.	1279 [3]	5.	482 [1]
2.	1271 [2]	6.	486 [2]
3.	1275 [3]	7.	485 [2]
4.	0.1277 [3]	8.	0.3333 [3]

Negative Numbers

Chapter 2

A negative number is a quantity with direction opposite to that of a positive number. When +5 ft is a distance in the upward direction, then −5 ft means the same distance but down instead of up. Or +5 ft can be to the right, and −5 ft to the left. Most important for electricity, the negative sign can represent an opposite polarity of voltage, or the opposite phase of 180°. In all cases, the opposite signs mean that the values oppose each other.

The topics in this chapter are:

2-1 Addition of Negative Numbers
2-2 Subtraction of Negative Numbers
2-3 Multiplication and Division of Negative Numbers

2-1 ADDITION OF NEGATIVE NUMBERS

For a negative number added to a positive number, the rule is, Take the difference between the two numbers and give the answer the sign of the larger number.

Example. Add 8 and −5.

Answer. $8 + (-5) = 3$

The 8 is the larger number. Therefore, the answer is positive.

Example. Add −8 and 5.

Answer. $(-8) + 5 = -3$

or $5 + (-8) = -3$

The 8 is negative. Therefore, the answer is negative.

When both numbers to be added are negative, just add the two numbers and give the answer the negative sign.

Example. $(-8) + (-3) = -11$

The 8 and 3 equals 11, but because both numbers added are negative the answer is −11.

Problems 2-A

Add these values:

1. $8 + 5 =$
2. $8 + (-5) =$
3. $(-5) + 8 =$
4. $(-5) + (-8) =$
5. $(-8) + 15 =$
6. $8 + (-15) =$
7. $17 + (-6) =$
8. $(-5) + (-3) =$

2-2 SUBTRACTION OF NEGATIVE NUMBERS

The rule is, Change the sign of the number to be subtracted (the *subtrahend*) and add that number to the other number. The addition is done by the rules just given in Sec. 2-1.

Example. Subtract −3 from 5. This is written as

$5 - (-3) = 5 + 3 = 8$

Example. Subtract -3 from -5. This is written as

$(-5) - (-3) = (-5) + 3 = -2$

Problems 2-B

Subtract these values:

1. $8 - (+5) =$
2. $8 - (-5) =$
3. $(-5) - (+8) =$
4. $(-5) - (-8) =$
5. $8 - (+15) =$
6. $8 - (-15) =$
7. $17 - (-6) =$
8. $(-12) - (-5) =$

2-3 MULTIPLICATION AND DIVISION OF NEGATIVE NUMBERS

For both multiplication and division the following rules apply:

1. When a negative number is multiplied or divided by a positive number, the answer is a *negative* number.
2. When two positive numbers are multiplied or divided, the answer is a *positive* number.
3. When two negative numbers are multiplied or divided, the answer is a *positive* number.

Examples.

$(6) \times (3) = 18$
$(6) \times (-3) = -18$
$(-6) \times (-3) = 18$
$(6) \div (3) = 2$
$(6) \div (-3) = -2$
$(-6) \div (-3) = 2$

When any *odd* number of negative values is multiplied or divided, the answer will be negative.

Example. $(-3) \times (-2) \times (-4) = -24$
This problem really combines the operations of

$(-3) \times (-2) = 6$

and $6 \times (-4) = -24$

Note that the two negative numbers are multiplied for a positive product. Then this positive 6 is multiplied by -4 for the answer of -24.

Problems 2-C

Do the following multiplications for a positive or negative answer:

1. $4 \times (-3) =$
2. $(-4) \times 3 =$
3. $(-3) \times (-3) =$
4. $7 \times (-2) =$
5. $3 \times 3 \times (-3) =$
6. $(-2) \times (-2) \times 5 =$
7. $3 \times (-4) =$
8. $(-6) \times 1/2 =$

Problems 2-D

Do the following divisions for a positive or negative answer:

1. $12 \div (-3) =$
2. $(-12) \div 3 =$
3. $(-12) \div (-3) =$
4. $8 \div (-4) =$
5. $12 \div 3 =$
6. $12 \div (-4) =$
7. $[8 \times (-4)] \times 2 =$
8. $4 \times \{[(-3) \times 6] \times (-2)\} =$

Problems 2-E

Do the following multiplications and divisions for a positive or negative answer:

1. $4 \times 2 \times (-3) =$
2. $(-4) \times 3 \times (-3) =$
3. $(-4) \times 2 \div (-8) =$
4. $4 \times 2 \times (-1) =$
5. $(-8) \times 6 \div (-2) =$
6. $(-6) \times (-6) =$
7. $(-3) \times (-3) \times (-3) =$
8. $8 \times (-3) \times 1 =$

Fractions

A fraction bar indicates the arithmetic operation of division. In the fraction $^2/_3$ the *numerator* 2 above the fraction bar is to be divided by the *denominator* 3 below. Such a fraction value is less than 1; that is, the numerator is less than the denominator. This is called a *proper fraction.* When the numerator of a fraction is greater than the denominator, the value of the fraction is more than 1 and it is called an *improper* fraction.

A proper fraction represents part of a complete unit. If the numerator of a fraction is held constant, as the denominator increases the value of the fraction will get smaller. Thus $^2/_5$ is less than $^2/_3$.

The topics in this chapter are:

3-1 Multiplication of Fractions
3-2 Division of Fractions
3-3 The Simplest Form of a Fraction
3-4 Multiplying or Dividing a Fraction by a Whole Number
3-5 Addition and Subtraction of Fractions
3-6 Negative Fractions
3-7 Reciprocals and Decimal Fractions
3-8 Working with Decimal Fractions

3-1 MULTIPLICATION OF FRACTIONS

The rule for the multiplication of fractions is, Multiply the numerators to obtain a new numerator and multiply the denominators to obtain a new denominator.

Example. Multiply $^2/_3$ by $^4/_7$.

Answer. $\frac{2}{3} \times \frac{4}{7} = \frac{2 \times 4}{3 \times 7} = \frac{8}{21}$

It may seem surprising that the product of two fractions less than 1 must be less than either of the two fractions. The reason is that this multiplication takes only a fractional part of the original fraction.

The fractions can be multiplied in any order.

Example. $\frac{2}{5} \times \frac{1}{3} = \frac{2}{15}$

or $\frac{1}{3} \times \frac{2}{5} = \frac{2}{15}$

Problems 3-A
Multiply the fractions.

1. $\frac{1}{3} \times \frac{2}{3} =$
2. $\frac{2}{7} \times \frac{1}{3} =$
3. $\frac{4}{9} \times \frac{2}{3} =$
4. $\frac{3}{4} \times \frac{3}{7} =$
5. $\frac{3}{5} \times \frac{2}{9} =$
6. $\frac{4}{7} \times \frac{2}{3} =$

3-2
DIVISION OF FRACTIONS

The rule for the division of fractions is, Invert the fraction that is the divisor and then multiply.

Example. Divide $^2/_5$ by $^5/_7$.

Answer. Write this as

$$\frac{2}{5} \div \frac{5}{7}$$

The divisor $^5/_7$ is inverted to $^7/_5$ and then multiplied by the other number.

$$\frac{2}{5} \times \frac{7}{5} = \frac{14}{25}$$

In division we must keep the fractions in the proper order. Thus $^2/_5$ divided by $^5/_7$ is not the same as $^5/_7$ divided by $^2/_5$. To prove this, we can do the problems both ways.

$$\frac{2}{5} \div \frac{5}{7} = \frac{2}{5} \times \frac{7}{5} = \frac{14}{25}$$

In the second case

$$\frac{5}{7} \div \frac{2}{5} = \frac{5}{7} \times \frac{5}{2} = \frac{25}{14}$$

which is the inverse of the correct answer.

Problems 3-B

Divide the fractions.

1. $\frac{1}{3} \div \frac{4}{5} =$ 3. $\frac{3}{8} \div \frac{8}{9} =$

2. $\frac{3}{7} \div \frac{2}{3} =$ 4. $\frac{3}{5} \div \frac{5}{7} =$

3-3
THE SIMPLEST FORM OF A FRACTION

A fraction is usually easiest to work with when the numerator and denominator have their lowest possible values. Consider, for instance, the fractions $^{50}/_{100}$, $^{12}/_{24}$, $^3/_6$, and $^{125}/_{250}$. Without actually realizing why, we may know that each of these numbers is equal to $^1/_2$. If we were going to use them in further calculations, it would be considerably easier to work with $^1/_2$ than $^{125}/_{250}$; yet the end result would be exactly the same.

Example. Multiply $^2/_9$ by $^{125}/_{250}$.

Answer. $\frac{2}{9} \times \frac{125}{250} = \frac{250}{2250} = \frac{1}{9}$

But $\frac{2}{9} \times \frac{1}{2} = \frac{2}{18} = \frac{1}{9}$

To reduce a fraction to simpler numbers, the numerator and denominator can be *divided* by the same number. This does not change the actual value of the fraction; it simply states the fraction in numbers easier to handle. In the fraction $^{250}/_{2250}$, both numerator and denominator were divided by 250. In the fraction $^2/_{18}$, both numerator and denominator were divided by 2.

Problems 3-C

Reduce to the lowest possible numerator and denominator.

1. $\frac{6}{9} =$ 3. $\frac{5}{20} =$

2. $\frac{6}{12} =$ 4. $\frac{9}{21} =$

Addition and subtraction of fractions often require finding a common denominator. This may involve *multiplying* numerator and denominator by the same number.

Example. Change $^2/_3$ to a fraction with a denominator of 18.

Answer. If we multiply the denominator by 6,

then the result will be 18. But we must also multiply the numerator by the same number to keep the value of the fraction constant. This gives us $2 \times 6 = 12$ for the new numerator. Or

$$\frac{2}{3} \times \frac{6}{6} = \frac{12}{18}$$

The fraction $^6/_6$ is, of course, equal to 1, and multiplying by 1 does not change the original number.

Problems 3-D

Raise each pair of fractions to the lowest common denominator.

1. $\frac{1}{2}$ and $\frac{1}{6}$
2. $\frac{1}{7}$ and $\frac{3}{14}$
3. $\frac{2}{5}$ and $\frac{3}{10}$
4. $\frac{2}{3}$ and $\frac{1}{6}$

Multiplication or division with fractions less than 1 can result in an answer of more than 1. Such a value as $^4/_3$ is an improper fraction. It can be converted to a mixed number, consisting of a whole number and a proper fraction. The method is to divide the numerator by the denominator. Then the answer (called the *quotient*) is a whole number with a remainder that is a proper fraction.

Example. Express the improper fraction $^5/_3$ as a mixed number.

Answer. $\frac{5}{3} = 5 \div 3 = 1\frac{2}{3}$

Problems 3-E

Change each improper fraction to a mixed number.

1. $\frac{7}{6} =$
2. $\frac{14}{3} =$
3. $\frac{6}{5} =$
4. $\frac{9}{4} =$

3-4 MULTIPLYING OR DIVIDING A FRACTION BY A WHOLE NUMBER

A whole number can be thought of as an improper fraction. For instance, 4 is the same as $^4/_1$.

Example. Multiply $^2/_3$ by 4.

Answer. $\frac{2}{3} \times 4 = \frac{2}{3} \times \frac{4}{1} = \frac{8}{3}$

Reduced to a mixed number, the product $^8/_3$ is equal to $2^2/_3$. Note that the procedure here is the same as multiplying only the numerator of the fraction by the whole number.

For division, the fractional divisor is inverted to multiply.

Example. Divide $^2/_3$ by 4.

Answer.

$$\frac{2}{3} \div 4 = \frac{2}{3} \div \frac{4}{1}$$
$$= \frac{2}{3} \times \frac{1}{4} = \frac{2}{12} \text{ or } \frac{1}{6}$$

Note that this method of dividing is the same as multiplying only the denominator of the proper fraction by the whole number.

Problems 3-F

Multiply or divide. Reduce each answer to lowest terms.

1. $\frac{1}{6} \times 3 =$
2. $\frac{3}{5} \times 5 =$
3. $\frac{2}{7} \times 3 =$
4. $\frac{1}{3} \div 3 =$
5. $\frac{3}{5} \div 5 =$
6. $\frac{2}{7} \div 3 =$

3-5
ADDITION AND SUBTRACTION OF FRACTIONS

Fractions must have the same denominator before they can be added or subtracted. Add or subtract the numerators of the fractions and put the result over the common denominator.

Examples. Add and subtract the following fractions as indicated: $3/7 + 2/7$; $3/7 - 2/7$.

Answer.

$$\frac{3}{7} + \frac{2}{7} = \frac{3+2}{7} = \frac{5}{7}$$

$$\frac{3}{7} - \frac{2}{7} = \frac{3-2}{7} = \frac{1}{7}$$

If the denominators of the fractions are not the same, they must be changed before adding or subtracting. To make the calculations as simple as possible, the *lowest common denominator* should be used.

Example. Add $5/12$ and $7/18$.

Answer. The example looks like this:

$$\frac{5}{12} + \frac{7}{18}$$

Since the denominators are not the same, the addition cannot be performed in this form. The multiples of 12 are 12, 24, 36, 48, etc. The multiples of 18 are 18, 36, 54, 72, etc. Since 36 is common to both denominators, each will be changed to 36:

$$\frac{5 \times 3}{12 \times 3} + \frac{7 \times 2}{18 \times 2}$$

or

$$\frac{15}{36} + \frac{14}{36}$$

When the denominators are the same, the numerators are simply added and the result is put over the denominator 36:

$$\frac{15 + 14}{36} = \frac{29}{36}$$

Therefore,

$$\frac{5}{12} + \frac{7}{18} = \frac{29}{36}$$

Problems 3-G

Add or subtract the fractions.

1. $\frac{4}{9} + \frac{1}{9} =$
2. $\frac{5}{9} + \frac{4}{9} =$
3. $\frac{7}{9} + \frac{5}{9} =$
4. $\frac{4}{9} - \frac{2}{9} =$
5. $\frac{5}{9} - \frac{4}{9} =$
6. $\frac{7}{9} - \frac{5}{9} + \frac{3}{9} =$

Problems 3-H

Combine the following fractions:

1. $\frac{4}{9} + \frac{1}{3} =$
2. $\frac{4}{9} - \frac{1}{3} =$
3. $\frac{3}{9} + \frac{2}{18} =$
4. $\frac{3}{4} - \frac{2}{5} =$
5. $\frac{1}{7} + \left(3 \times \frac{2}{7}\right) =$
6. $2 + \left(4 \times \frac{1}{2}\right) =$
7. $\left(\frac{3}{7} \times \frac{2}{7}\right) + \frac{5}{49} =$
8. $\frac{5}{6} - \frac{1}{3} - \frac{1}{2} =$
9. $\left(\frac{3}{5} \times 2\right) - \frac{3}{5} =$
10. $\left(\frac{1}{3} \times 3\right) - \frac{1}{7} =$

3-6
NEGATIVE FRACTIONS

A minus sign in front of a fraction bar makes the entire fraction negative. Use such a fraction as a negative number.

Example. Add $(-1/5)$ to $3/5$. This can be written as

$$\frac{3}{5} + \left(-\frac{1}{5}\right) = \frac{3}{5} - \frac{1}{5} = \frac{2}{5}$$

Example. Subtract $(-1/5)$ from $3/5$. This can be written as

$$\frac{3}{5} - \left(-\frac{1}{5}\right) = \frac{3}{5} + \frac{1}{5} = \frac{4}{5}$$

For multiplication and division, remember that a minus sign in either the numerator or denominator makes the fraction negative.

Examples. $\frac{3}{5} \times \left(\frac{-1}{5}\right) = \frac{-3}{25} = -\frac{3}{25}$

$\frac{3}{5} \times \left(\frac{1}{-5}\right) = \frac{3}{-25} = -\frac{3}{25}$

However, a minus sign in both the numerator and denominator means the fraction is really positive. The reason is that division of the two negative numbers in the numerator and denominator would result in a positive quotient.

Example. $\frac{3}{5} \times \left(\frac{-1}{-5}\right) = \frac{3}{5} \times \frac{1}{5} = \frac{3}{25}$

Problems 3-I

Combine these fractions.

1. $\frac{4}{9} - \left(-\frac{2}{9}\right) =$
2. $\frac{4}{9} \times \left(\frac{-2}{9}\right) =$
3. $\frac{4}{9} \div \left(\frac{-2}{9}\right) =$
4. $\frac{4}{9} \times \left(\frac{-2}{-9}\right) =$

3-7 RECIPROCALS AND DECIMAL FRACTIONS

A reciprocal of any number is 1 divided by that number. For instance, the reciprocal of 7 is $1/7$. Furthermore, the reciprocal of $1/7$ is equal to 7. The reason is that

$$1 \div \frac{1}{7} = 1 \times \frac{7}{1} = 7$$

The reciprocals of the digits 2 to 9 often must be converted to decimal numbers. These values are as follows:

$1/2 = 0.5$ $\quad 1/6 = 0.166 \cdots$
$1/3 = 0.333 \cdots$ $\quad 1/7 = 0.142857 \cdots$
$1/4 = 0.25$ $\quad 1/8 = 0.125$
$1/5 = 0.2$ $\quad 1/9 = 0.111 \cdots$

The decimal values for $1/3$, $1/6$, $1/7$, and $1/9$ are inexact and can be rounded off as explained in Chap. 1. The values for $1/2$, $1/4$, $1/5$, and $1/8$ are exact to the number of places shown.

The decimal equivalents for each of these fractions result from just dividing each denominator into the numerator 1. Note also that $1/1 = 1$ and $0/1 = 0$. However, the division $1/0$ is indeterminate. Because it is extremely large without limits, it is generally called infinity and given the symbol ∞.

A special reciprocal worth memorizing is $1/(2\pi)$, because this factor is often used in electronics formulas. The value of the constant π rounded off to the nearest hundredth is 3.14, and $2\pi = 6.28$. The reciprocal, then, is

$$\frac{1}{2\pi} = \frac{1}{6.28} = 0.159$$

Problems 3-J

Give the reciprocal as a decimal value.

1. 5
2. 50
3. $\sqrt{25}$
4. $1/2$
5. 6 + 3
6. 2π

3-8 WORKING WITH DECIMAL FRACTIONS

A decimal fraction does not have a fraction bar. An example is 0.5, which is the same as $1/2$. For this reason decimal fractions are often easier to work with; they do not have numerators and denominators. However, you must keep track of the decimal point.

Example. Add 0.5, 0.2, and 0.1.

Answer. This problem can be shown as

$$\begin{array}{r} 0.5 \\ 0.2 \\ \underline{0.1} \\ 0.8 \end{array}$$

For addition or subtraction, the decimal points are in line for all the numbers.

Multiplication and division of decimal fractions is the same as with whole numbers.

Example. Multiply 0.4×0.2.

Answer. This problem can be shown as

$$\begin{array}{r} 0.4 \\ \underline{\times 0.2} \\ 0.08 \end{array}$$

There are two decimal places in the product, because each of the two factors has one decimal place.

Example. Divide 0.6 by 0.2.

Answer. This problem can be shown as

$$0.2\overline{)0.6} = 2\overline{)6} = 3$$

The decimal points are moved to make the divisor a whole number.

Problems 3-K

Solve the following:

1. $0.2 + 0.3 + 0.4 =$
2. $0.2 + 0.3 - 0.1 =$
3. $0.2 + 0.05 =$
4. $0.2 \times 0.3 =$
5. $0.2 \times (-0.3) =$
6. $0.3 \times 0.02 =$

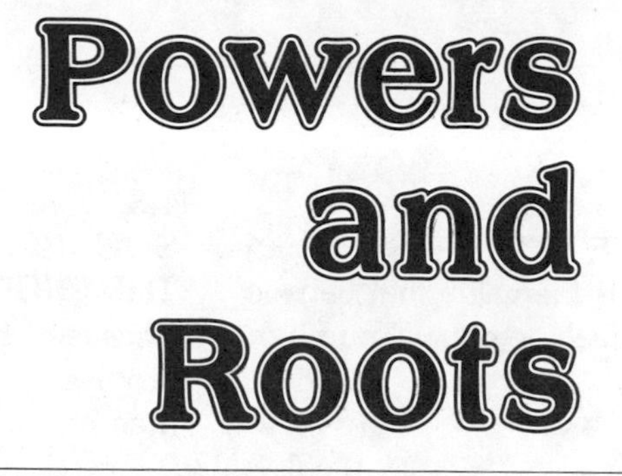

Multiplying a number by itself is the same as raising the number to a higher power. The *exponent* written above the number indicates how many times the number is used as a factor in multiplication. For instance, 5×5 is written 5^2. The 5 is the *base* number raised to the second power by the exponent 2. The cube is the third power of a number. Any base number can be raised to any power.

Example. Write the following with exponents.

$2 \times 2 \times 2 \times 2 = 2^4 = 16$

$5 \times 5 \times 5 = 5^3 = 125$

The topics in this chapter are:

4-1
POSITIVE EXPONENTS

The purpose of using positive exponents is to provide a shortcut method of indicating repeated multiplications of the same number. Exponents are especially useful for powers of 10, as explained in Chap. 5.

Problems 4-A

Raise the base number to the power indicated by the exponent.

1. $2^3 =$
2. $3^2 =$
3. $10^4 =$
4. $5^3 =$
5. $6^2 =$
6. $4^3 =$
7. $3^3 =$
8. $3^4 =$

4-2
ROOTS OF POSITIVE NUMBERS

The *root* of a number is the value that can be multiplied by itself to equal the original number.

Example. Find the cube root of 125. This can be written as

$$\sqrt[3]{125}$$

Answer. Although there are arithmetic procedures that can be used to find roots, most often either a table of roots or an electronic calculator with this capability is used. For certain commonly used numbers, the square roots and cube roots are memorized. For the above example we find

$$5 \times 5 \times 5 = 125$$

Therefore,

$$\sqrt[3]{125} = 5$$

Examples. Find each of the following roots.

$\sqrt{4} = 2$ [and -2, since $(-2) \times (-2) = 4$]
$\sqrt[3]{8} = 2$ [but *not* -2, since $(-2) \times (-2) \times (-2) = -8$]
$\sqrt[4]{16} = 2$ (Again, -2 is also an answer.)
$\sqrt[5]{32} = 2$

Problems 4-B

Find the value of the indicated root.

1. $\sqrt[3]{8} =$
2. $\sqrt{9} =$
3. $\sqrt[4]{10{,}000} =$
4. $\sqrt[3]{125} =$
5. $\sqrt{36} =$
6. $\sqrt[3]{64} =$
7. $\sqrt{64} =$
8. $\sqrt[4]{16} =$

The radical sign $\sqrt{}$ is generally used with an index number to indicate the root. When no index number is shown, it is assumed to be 2 for the square root.

4-3
SQUARES AND ROOTS FOR THE DIGITS

Because they occur so often in numerical problems, the squares of the digits should be memorized. These are as follows:

$0^2 = 0 \times 0 = 0$	$5^2 = 5 \times 5 = 25$
$1^2 = 1 \times 1 = 1$	$6^2 = 6 \times 6 = 36$
$2^2 = 2 \times 2 = 4$	$7^2 = 7 \times 7 = 49$
$3^2 = 3 \times 3 = 9$	$8^2 = 8 \times 8 = 64$
$4^2 = 4 \times 4 = 16$	$9^2 = 9 \times 9 = 81$

It should be noted that any power or root of zero is still zero. Also, any power or root of 1 is still 1.

The square of a number is that number multiplied by itself. It does *not* mean double the number. It happens that $2^2 = 4$, which is also double the 2, but this is true only with the digit 2. The square of 3 is 9, but doubling 3 would give 6.

Problems 4-C

Give the square of the following:

1. $5^2 =$
2. $3^2 =$
3. $1^2 =$
4. $7^2 =$
5. $10^2 =$
6. $9^2 =$

The squares of the digits should be learned forward and backward for some important square roots also. Since $5^2 = 25$, then the square root of 25 is 5, as $5 \times 5 = 25$. Some common square root values that should be memorized are the following:

$\sqrt{1} = 1$	$\sqrt{25} = 5$
$\sqrt{2} = 1.414$	$\sqrt{36} = 6$
$\sqrt{3} = 1.732$	$\sqrt{49} = 7$
$\sqrt{4} = 2$	$\sqrt{64} = 8$
$\sqrt{9} = 3$	$\sqrt{81} = 9$
$\sqrt{16} = 4$	$\sqrt{100} = 10$

The reason why the squares of the digits and their roots are used so often is the fact that very large or small numbers can be converted to these values as a factor with the appropriate multiple of 10.

Example. Express 400 as a factor with a multiple of 10.

Answer. $400 = 4 \times 100 = 4 \times 10^2$

Problems 4-D

Give the following square roots. Check your answer by multiplying.

1. $\sqrt{25} =$ 4. $\sqrt{100} =$
2. $\sqrt{36} =$ 5. $\sqrt{9} =$
3. $\sqrt{81} =$ 6. $\sqrt{4} =$

4-4 POWERS OF A NEGATIVE NUMBER

Since raising a number to a power is the same as repeated multiplication, the rules for multiplying negative numbers apply here.

Raising a negative number to an even power, for example, squaring a negative number, results in a positive answer.

Examples. $(-3)^2 = (-3) \times (-3) = 9$
$(-2)^4 = (-2) \times (-2) \times (-2) \times (-2) = 16$

Raising a negative number to an odd power, for example, cubing a negative number, results in a negative answer.

Examples. $(-2)^3 = (-2) \times (-2) \times (-2) = -8$
$(-3)^5 = (-3) \times (-3) \times (-3) \times (-3) \times (-3) = -243$

Problems 4-E

Raise these numbers to the power indicated:

1. $(3)^2 =$ 5. $(-1)^6 =$
2. $(-3)^2 =$ 6. $(+1)^6 =$
3. $(2)^3 =$ 7. $(-4)^2 =$
4. $(-1)^5 =$ 8. $(-4)^3 =$

4-5 ROOTS OF A NEGATIVE NUMBER

Since an even power of a negative number always leads to a positive number answer, it is not possible to work back from a negative number to a positive even root. However, it is possible to find odd roots of negative numbers.

Example. Find the cube root of -8.

Answer. $\sqrt[3]{-8} = -2$

To prove the answer we need only find the *cube* of -2.

$$(-2)^3 = (-2) \times (-2) \times (-2) = -8$$

Though we cannot calculate the square root of a negative number by arithmetic, such numbers are useful in electricity and electronics. To get around this problem we *assume* that square roots do exist for negative numbers. These roots are called *imaginary numbers.* Since the positive and negative numbers we have been dealing with up to this point can be represented on the horizontal axis, the imaginary numbers are indicated on a vertical axis, called the *j axis.* In order to indicate that a number is imaginary, it is preceded by the letter j, representing $\sqrt{-1}$.

Example. Represent $\sqrt{-4}$ as an imaginary number.

Answer. $\sqrt{-4} = \sqrt{-1} \times \sqrt{4} = j\sqrt{4} = j2$

Since j preceding a number indicates that the number is to be rotated to the vertical j axis, we often use the expression *j operator.** The *j* operator really represents the angle of 90°.

Problems 4-F
Find the indicated root.

1. $\sqrt[3]{8} =$ 5. $\sqrt{-64} =$
2. $\sqrt[3]{-8} =$ 6. $\sqrt[3]{-64} =$
3. $\sqrt[3]{-27} =$ 7. $\sqrt[3]{125} =$
4. $\sqrt{25} =$ 8. $\sqrt[4]{16} =$

Another possibility is to have a number that is positive but with a negative exponent, as in 10^{-2}. The negative exponent, however, is only a method of indicating a reciprocal. For instance, 5^{-2} is the reciprocal of 5^2. And 3^{-2} is the reciprocal of 3^2. These values can be stated as follows:

$$5^2 = 25$$

$$5^{-2} = \frac{1}{5^2} = \frac{1}{25}$$

And

$$3^2 = 9$$

$$3^{-2} = \frac{1}{3^2} = \frac{1}{9}$$

Note that these reciprocal values are positive numbers. The negative exponents for reciprocals are especially useful with powers of 10 to indicate decimal fractions. This application is explained in detail in the next chapter.

4-6 POWERS AND ROOTS OF FRACTIONS

When a fraction is raised to a power, both the numerator and denominator must be raised to that power.

Example. Cube $^2/_3$.

Answer. The numerator must be cubed, and the denominator must also be cubed.

$$\left(\frac{2}{3}\right)^3 = \frac{2 \times 2 \times 2}{3 \times 3 \times 3} = \frac{8}{27}$$

To find the root of a fraction the reverse process is used in that the root of both the numerator and the denominator must be found.

Example. Find $\sqrt{^8/_{27}}$.

Answer. $\sqrt{\frac{8}{27}} = \frac{\sqrt{8}}{\sqrt{27}} = \frac{2}{3}$

This answer, when cubed, will result in the original fraction.

Problems 4-G
Find the powers or roots of the following fractions:

1. $(^2/_3)^2 =$ 5. $\sqrt{^4/_9} =$
2. $(^3/_7)^2 =$ 6. $\sqrt{^9/_{49}} =$
3. $(^1/_2)^4 =$ 7. $\sqrt[4]{^1/_{16}} =$
4. $(^1/_3)^2 =$ 8. $\sqrt{^1/_9} =$

An interesting fact about a proper fraction is that raising it to a power makes the fraction smaller. For instance, $(^1/_2)^2 = {^1/_4}$. The answer of $^1/_4$ is smaller than $^1/_2$ because the denominator is larger with the same numerator. Remember that a proper fraction has a value less than 1 because the denominator is larger than the numerator. Raising the fraction to a power accentuates this property by increasing the larger number in the denominator more than the increase in the numerator.

Problems 4-H
Pick out the larger value in each of the following pairs of fractions:

* For details of the *j* operator, See Bernard Grob, Basic Electronics, 4th ed., Chap. 26, McGraw-Hill Book Company, New York, 1977.

1. $^1/_2$ or $^1/_4$
2. $^2/_3$ or $^4/_9$
3. $^1/_5$ or $^1/_{25}$
4. $^9/_{49}$ or $^3/_7$
5. $^1/_{100}$ or $^1/_{10}$
6. 10 or $^1/_{10}$

4-7 POWERS AND ROOTS OF NUMBERS WITH EXPONENTS

Very often, it is necessary to raise to a power or find the root of a number that already has an exponent, as in $(2^2)^3$. To perform this operation the rule is, Multiply the exponents and make the product the new exponent for the original base number.

Example. Find the value of $(4^2)^3$.

Answer. The product of the exponents is

$2 \times 3 = 6$

This 6 is the new exponent for the base 4:

4^6

The value of 4^6 is equal to

$4 \times 4 \times 4 \times 4 \times 4 \times 4 = 4096$

This value is the same as $(16)^3$:

$16 \times 16 \times 16 = 4096$

A similar but opposite process is used for finding roots. The rule is, Divide the original exponent by the root and use the answer as the new exponent of the original base.

Example. Find $\sqrt[3]{8^6}$.

Answer. Divide the exponent 6 by the root 3. The result is

$\frac{6}{3} = 2$

The 2 is the new exponent for the original base 8. Thus

$\sqrt[3]{8^6} = 8^2 = 64$

Problems 4-I

Find the powers and roots, as indicated.

1. $(9^2)^2 =$
2. $[(-3)^2]^3 =$
3. $\sqrt{7^2} =$
4. $(4 \times 2^2)^2 =$
5. $\sqrt[4]{8^4} =$
6. $\sqrt[3]{10^6} =$
7. $\sqrt[3]{5^6} =$
8. $\sqrt{36 \times 8^4} =$

4-8 SQUARES AND ROOTS WITH FACTORS

Factors are parts of numbers which, when multiplied together, produce the number. In $2 \times 4 = 8$, the 2 and 4 are factors of 8.

In squaring or taking the square root, the operation can be applied to each of the factors separately.

Examples. Find the following square roots, using factors:

$$\sqrt{25 \times 9} = \sqrt{25} \times \sqrt{9} = 5 \times 3 = 15$$
$$\sqrt{49 \times 4} = \sqrt{49} \times \sqrt{4} = 7 \times 2 = 14$$
$$\sqrt{49 \times 25} = \sqrt{49} \times \sqrt{25} = 7 \times 5 = 35$$

Examples. Find the following squares, using factors:

$$(2 \times 3)^2 = (2)^2 \times (3)^2 = 4 \times 9 = 36$$
$$(4 \times 5)^2 = 4^2 \times 5^2 = 16 \times 25 = 400$$

This procedure of separating the factors can be used for any power or root, but the examples here are for the common problem of finding a square or square root.

Problems 4-J

Find the square or square root of the following:

1. $(4 \times 2)^2 =$
2. $(9 \times 3^2)^2 =$
3. $(5^2 \times 3^2)^2 =$
4. $(7 \times 10^4)^2 =$
5. $\sqrt{16 \times 4} =$
6. $\sqrt{81^4} =$
7. $\sqrt{2^4 \times 3^2} =$
8. $\sqrt{49 \times 10^8} =$

4-9
SQUARES AND ROOTS WITH TERMS

Terms are numbers in a group that are to be added or subtracted. For instance, in $(2 + 7)$, the 2 and 7 are terms. To find the square or root, all the terms must be combined first.

Example. Square $(2 + 7)$, or find $(2 + 7)^2$.

Answer. This procedure is different from the method with factors. If you square each term separately, the answer will be wrong. First combine the terms by adding: $2 + 7 = 9$. Then square the sum: $(9)^2 = 81$. Therefore, $(2 + 7)^2 = 81$.

Note that $(2 + 7)^2$ is not equal to $2^2 + 7^2$, as $4 + 49 = 53$.

Problems 4-K

Solve the following:

1. $(3 + 4)^2 =$
2. $[(3 \times 10^3) + (4 \times 10^3)]^2 =$
3. $(3 - 4)^2 =$
4. $[(2 \times 10^{-4}) + (3 \times 10^{-4})]^2 =$
5. $(2 + 4)^2 =$
6. $(3 + 7 + 9)^2 =$

The same procedure applies to the square root for a group of terms. They must all be combined before the root is found. For instance, $\sqrt{16 + 9} = \sqrt{25} = 5$. If you take the roots separately, the answer of $4 + 3 = 7$ will be wrong. These are terms, not factors. A power or root for factors can be applied separately, but terms must be combined first.

Example. Find $\sqrt{72 - 8}$.

Answer. Combine terms first: $72 - 8 = 64$. Then find the root: $\sqrt{64} = 8$.

This procedure of combining terms before you find the square or square root also applies for any power or root.

Problems 4-L

Solve the following:

1. $\sqrt{16 + 9} =$
2. $\sqrt{35 + 1} =$
3. $\sqrt{5^2 - 4^2} =$
4. $\sqrt{(7 \times 10^6) + (2 \times 10^6)} =$
5. $\sqrt{5^2 - 3^2} =$
6. $\sqrt{12 + 7 + 6} =$

As an application of these methods, an important formula in electronics is $Z = \sqrt{R^2 + X^2}$, where Z is the impedance, R the resistance, and X the reactance in the circuit, all in ohms units. To calculate Z, the R and X within the square root sign must be evaluated and combined before you take the square root of the grouping.

Example. If $R = 3$ ohms and $X = 4$ ohms, find Z.

Answer.

$$Z = \sqrt{R^2 + X^2} = \sqrt{3^2 + 4^2} = \sqrt{9 + 16} = \sqrt{25}$$

$$Z = 5 \text{ ohms}$$

Problems 4-M

With the formula $Z = \sqrt{R^2 + X^2}$, find Z for the following values of R and X:

1. $R = 3, X = 4$
2. $R = 4, X = 3$
3. $R = 4, X = 4$
4. $R = 3, X = -4$
5. $R = 6, X = -6$
6. $R = 4, X = 8$

Problems 4-N

With the formula $X = \sqrt{Z^2 - R^2}$, find X for the following values of Z and R:

1. $Z = 5, R = 4$
2. $Z = 14.14, R = 10$
3. $Z = 8.48, R = 6$
4. $Z = 6, R = 2$
5. $Z = 8, R = 4$
6. $Z = 17, R = 9$

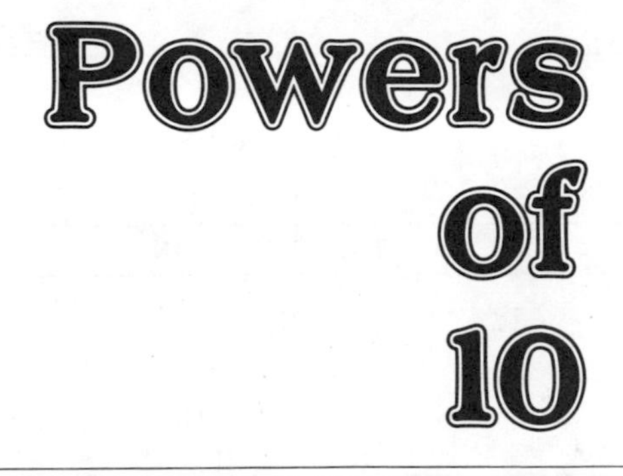

From Chap. 4 we learned that a power or exponent is written above a number to indicate how many times the number is used as a factor in multiplication by itself. As an example, 10^3 is the same as $10 \times 10 \times 10$, which equals 1000. The exponent here is 3, and 10 is the *base* for the exponent. The base is raised to a power indicated by the exponent.

The base 10 is common for exponents, because 10 is the basis of decimal numbers for counting and for the decimal multiples in the metric system of units. In general, powers of 10 help keep track of the decimal point in arithmetical operations involving very large or very small numbers.

The topics in this chapter are:

5-1 POSITIVE EXPONENTS OF 10

Numbers greater than 1 can be written as powers of 10 by using positive exponents.

Note that 10 and 10^1 are the same. A number written without a power is assumed to have the exponent 1.

Note also that 10^0 has the value of 1. Any base to the zero power is equal to unity, because it represents a fraction with the same numerator and denominator.

A higher positive exponent means a larger number. For instance, 10^3 for 1000 is more than 10^2 for 100.

Example. Represent 1, 10, 100, 1000, 10,000, 100,000, and 1,000,000 using powers of 10.

Answer.

POWER OF 10	MULTIPLICATION	PRODUCT NUMBER
10^0	—	1
10^1	10	10
10^2	10×10	100
10^3	$10 \times 10 \times 10$	1000
10^4	$10 \times 10 \times 10 \times 10$	10,000
10^5	$10 \times 10 \times 10 \times 10 \times 10$	100,000
10^6	$10 \times 10 \times 10 \times 10 \times 10 \times 10$	1,000,000

Problems 5-A

Convert to powers of 10.

1. 100 = 4. 100,000 =
2. 1000 = 5. 1,000,000 =
3. 10,000 = 6. 10,000,000 =

Problems 5-B

Write the following as common numbers:

1. 10^3 = 3. 10^6 =
2. 10^2 = 4. 10^4 =

5-2 NEGATIVE EXPONENTS OF 10

Numbers less than 1 can also be written as powers of 10. In this case a negative exponent must be used. A negative exponent of 10 shows powers of tenths, compared to 10 for a positive exponent.

Example. Write 0.1, 0.01, 0.001, 0.0001, 0.00001, and 0.000001 using powers of 10.

Answer.

$$10^{-1} = {}^1/_{10} = 0.1$$
$$10^{-2} = {}^1/_{10} \times {}^1/_{10} = {}^1/_{100} = 0.01$$
$$10^{-3} = {}^1/_{10} \times {}^1/_{10} \times {}^1/_{10} = {}^1/_{1000} = 0.001$$
$$10^{-4} = {}^1/_{10} \times {}^1/_{10} \times {}^1/_{10} \times {}^1/_{10} = {}^1/_{10,000} = 0.0001$$
$$10^{-5} = {}^1/_{10} \times {}^1/_{10} \times {}^1/_{10} \times {}^1/_{10} \times {}^1/_{10} = {}^1/_{100,000} = 0.00001$$
$$10^{-6} = {}^1/_{10} \times {}^1/_{10} \times {}^1/_{10} \times {}^1/_{10} \times {}^1/_{10} \times {}^1/_{10} = {}^1/_{1,000,000} = 0.000\,001$$

Problems 5-C

Convert to powers of 10.

1. 0.001 = 4. 100 =
2. $\frac{1}{1000}$ = 5. $\frac{1}{1,000,000}$ =
3. 0.01 = 6. 0.000 001 =

Problems 5-D

Convert to decimal fractions.

1. 10^{-3} = 3. 10^{-2} =
2. 10^{-1} = 4. 10^{-6} =

Problems 5-E

Convert to proper fractions.

1. 10^{-3} = 3. 10^{-2} =
2. 10^{-1} = 4. 10^{-6} =

Notice that the larger the negative exponent the smaller the number. For instance, 10^{-3} is smaller than 10^{-2}, as ${}^1/_{1000}$ is less than ${}^1/_{100}$. When you increase the negative exponent by 1, the number becomes 10 times smaller. To go the other way, decrease a negative exponent by 1 to make the number 10 times larger. These rules for negative exponents are opposite from the rules for positive exponents.

Problems 5-F

Change the exponent to make the following numbers 10 times larger:

1. 10^3 = 3. 10^{-6} =
2. 10^{-3} = 4. 10^6 =

5. $10^7 =$ 7. $10^{-5} =$
6. $10^{-7} =$ 8. $10^5 =$

Problems 5-G

Change the exponent to make the following numbers 10 times smaller:

1. $10^3 =$ 5. $10^7 =$
2. $10^{-3} =$ 6. $10^{-7} =$
3. $10^{-6} =$ 7. $10^{-5} =$
4. $10^6 =$ 8. $10^5 =$

5-3 CONVERTING TO POWERS OF 10

All the numbers used so far in this chapter have been perfect powers of 10. However, use of the powers of 10 is not limited to these numbers. The general procedure for using powers of 10 with any number is to convert the given number into two factors, where one factor (the power of 10) is used merely to place the decimal point.

Example. Write 750 as the product of two numbers, one of which is a power of 10.

Answer.
$$75 \times 10 = 750$$
$$7.5 \times 10^2 = 750$$
$$0.75 \times 10^3 = 750$$

Notice that the power of 10 merely positions the decimal point in the original number. The exponent equals the number of places the decimal point is moved. What we have then is another way of writing a number so that the original number is easier to handle. In other words, we have a shorthand way of writing otherwise long numbers.

Example. Write 1,640,000 using powers of 10 for millions.

Answer.
$$1{,}640{,}000 = 1.64 \times 1{,}000{,}000 = 1.64 \times 10^6$$

Example. Write each of the following as a number between 1 and 10 multiplied by a power of 10: 100; 980; 9800; 9840.

Answer.
$$100 = 1 \times 100 = 1 \times 10^2$$
$$980 = 9.8 \times 100 = 9.8 \times 10^2$$
$$9800 = 9.8 \times 1000 = 9.8 \times 10^3$$
$$9840 = 9.84 \times 1000 = 9.84 \times 10^3$$

In the above examples only positive exponents were used, because the numbers are greater than 1. However, the same procedure applies to decimal fractions less than 1. In this case, negative exponents are used to mark off the number of places the decimal point is moved for multiples of tenths.

Example. Write each of the following as a number between 1 and 10 multiplied by a power of 10: 0.01; 0.05; 0.053.

Answer.
$$0.01 = 1 \times 0.01 = 1 \times 10^{-2}$$
$$0.05 = 5 \times 0.01 = 5 \times 10^{-2}$$
$$0.053 = 5.3 \times 0.01 = 5.3 \times 10^{-2}$$

Moving the point to the right multiplies the value for a coefficient greater than 1. However, the negative exponent is a division for the same number of places, to keep the number the same. In short, 0.05 and 5.0×10^{-2} are two ways to write the same value.

The Coefficient of the Base. In a number like 9×10^3, the 9 is a coefficient of base 10 with its power. The coefficient is a factor to be multiplied. In this example the coefficient 9 and the base number 10^3, or 1000, are multiplied for the product: $9 \times 1000 = 9000$.

When no coefficient is given, it is assumed to be 1. For instance, 1000 can be written as 10^3. Also, 0.01 is 10^{-2}.

The purpose of having the coefficient is to factor out the part of the number that is not a perfect power of 10. In this way all numbers can be written as powers of 10 with the appropriate coefficient.

Scientific or Engineering Notation. Numbers that are written in the form of a power of 10 and a coefficient between 1 and 10 are said to be written in scientific or engineering notation. For instance, 4.5×10^3 is in this notation. Actually, 45×10^2 is the same value of 4500, but the scientific notation is usually preferable in electrical and electronics work. One reason is that arithmetical calculations with the coefficient are easier with numbers between 1 and 10. Another feature of this form is that when logarithms are used, the power of 10 is the characteristic in the common logarithm of the number.

Problems 5-H

Convert to powers of 10, in scientific notation

1.	400 =	6.	0.04 =
2.	470 =	7.	0.047 =
3.	4000 =	8.	0.004 =
4.	4700 =	9.	0.0047 =
5.	8,000,000 =	10.	0.000 008 =

Problems 5-I

Convert each of the following to a number without the power of 10 notation:

1.	$4 \times 10^2 =$	6.	$4 \times 10^{-2} =$
2.	$4.7 \times 10^2 =$	7.	$4.7 \times 10^{-2} =$
3.	$4 \times 10^3 =$	8.	$4 \times 10^{-3} =$
4.	$4.7 \times 10^3 =$	9.	$4.7 \times 10^{-3} =$
5.	$8 \times 10^6 =$	10.	$8 \times 10^{-6} =$

The scientific notation is not always necessary. In expressing the final answer to a problem, it may be better to use the third or sixth power in order to have units often used with metric prefixes. For instance, 18×10^3 V can be used for 18,000 volts rather than 1.8×10^4 V. The prefix *kilo* means 1000, or 10^3. Thus 18 kilovolts (kV) = 18×10^3 V.

Example. Express 0.042 amperes (A) in terms of milliamperes (mA).

Answer. The prefix *milli* means 1/1000 or 0.001.

Since $10^{-3} = 0.001$, we can express milliamperes in terms of 10^{-3}.

$$0.042 \text{ A} = 42 \times 10^{-3} \text{ A} = 42 \text{ mA}$$

Problems 5-J

Convert to a power of 10 with an exponent of 3 or 6, plus or minus.

1.	$1.8 \times 10^4 =$	5.	$20 \times 10^2 =$
2.	$4.2 \times 10^{-2} =$	6.	$20 \times 10^5 =$
3.	$80 \times 10^5 =$	7.	$20 \times 10^{-4} =$
4.	$76 \times 10^2 =$	8.	$20 \times 10^{-7} =$

5-4 METRIC PREFIXES

Probably the most important use of powers of 10 is to express metric prefixes. Two common examples are 10^3 for *kilo,* or 1000, and 10^6 for *mega,* or 1,000,000. The corresponding fractional values are 10^{-3}, or 0.001, for *milli* and 10^{-6}, or 0.000 001, for *micro.*

These metric powers of 10 are often used in electronics with the units of volts (V), amperes (A) of current, watts (W) of power, ohms (Ω) of resistance, seconds (s) of time, hertz (Hz) of frequency, henrys (H) of inductance, and farads (F) of capacitance. See Table 5-1 for a complete listing of the metric prefixes from 10^{12} to 10^{-12}, including examples with units. The metric prefixes actually specify decimal multiples and submultiples.

The following examples show how electrical units can be written with powers of 10 for metric prefixes:

$$9 \text{ kilovolts} = 9 \text{ kV} = 9 \times 10^3 \text{ V}$$
$$4 \text{ milliamperes} = 4 \text{ mA} = 4 \times 10^{-3} \text{ A}$$
$$5 \text{ megohms} = 5 \text{ M}\Omega = 5 \times 10^6 \ \Omega$$
$$6 \text{ microseconds} = 6 \ \mu\text{s} = 6 \times 10^{-6} \text{ s}$$

Note the important difference between capital M for 10^6 and small m for 10^{-3}. The small k is used

TABLE 5-1. Metric Prefixes*

PREFIX	SYMBOL	POWER OF 10	VALUE	EXAMPLE
tera	T	10^{12}	1,000,000,000,000	2 THz = 2×10^{12} Hz
giga	G	10^{9}	1,000,000,000	9 GHz = 9×10^{9} Hz
mega	M	10^{6}	1,000,000	5 MΩ = 5×10^{6} Ω
kilo	k	10^{3}	1000	9 kV = 9×10^{3} V
hecto	h	10^{2}	100	8 hm = 8×10^{2} m†
deka	da	10^{1}	10	6 dam = 6×10 m†
deci	d	10^{-1}	0.1	7 dm = 7×10^{-1} m†
centi	c	10^{-2}	0.01	4 cm = 4×10^{-2} m†
milli	m	10^{-3}	0.001	8 mA = 8×10^{-3} A
micro	μ	10^{-6}	0.000 001	5 μm = 5×10^{-6} m†
nano	n	10^{-9}	0.000 000 001	3 ns = 3×10^{-9} s
pico	p	10^{-12}	0.000 000 000 001	5 pF = 5×10^{-12} F

* The complete list of metric prefixes also includes peta = 10^{15}, exa = 10^{18}, femto = 10^{-15}, and atto = 10^{-18}.
† The m is for meter, a length equal to 39.37 in. The μm was formerly called a *micron*.

for kilo, as capital K is reserved for temperature units in kelvins.

Problems 5-K
Change to powers of 10.

1. 5 μV =
2. 12 kV =
3. 47 pF =
4. 16 mA =
5. 42 mH =
6. 6.8 MΩ =
7. 12 μs =
8. 25 mW =

Problems 5-L
Change to units with a metric prefix instead of the power of 10.

1. 5×10^{-6} V =
2. 12×10^{3} V =
3. 47×10^{-12} F =
4. 16×10^{-3} A =
5. 42×10^{-3} H =
6. 6.8×10^{6} Ω =
7. 12×10^{-6} s =
8. 25×10^{-3} W =

5-5 MULTIPLICATION WITH POWERS OF 10

To multiply numbers made up of powers of 10, multiply the coefficients to obtain the new coefficient and add the exponents to obtain the new power of 10.

Example. Multiply 200 by 40,000.

Answer. $200 = 2 \times 10^{2}$
$40{,}000 = 4 \times 10^{4}$

Then the problem is

$(2 \times 10^{2}) \times (4 \times 10^{4})$

$2 \times 4 = 8$ for the coefficients
$2 + 4 = 6$ for the exponents

Finally,

$(2 \times 10^{2}) \times (4 \times 10^{4}) = 8 \times 10^{6}$

or

$$200 \times 40{,}000 = 8{,}000{,}000$$

The reason why the exponents are added for multiplication of the base is that each increase of one in a positive exponent is equivalent to moving the decimal point one place, as when multiplying by 10.

Problems 5-M

Multiply in powers of 10.

1. $(2 \times 10^2) \times (3 \times 10^4) =$
2. $(4 \times 10) \times (2 \times 10) =$
3. $(7 \times 10^7) \times (1 \times 10) =$
4. $(3 \times 10^7) \times (2 \times 10^8) =$
5. $(5 \times 10^5) \times (1 \times 10^2) =$
6. $(3 \times 10^3) \times (2 \times 10^2) =$
7. $(7 \times 10^2) \times (1 \times 10^2) =$
8. $(2.5 \times 10^4) \times (2 \times 10^4) =$

When there are negative exponents, they are also added. The new exponent is a larger negative number.

Example. Multiply 0.02 by 0.1.

Answer. $0.02 = 2 \times 10^{-2}$
$0.1 = 1 \times 10^{-1}$

Then the problem is

$(2 \times 10^{-2}) \times (1 \times 10^{-1})$

$1 \times 2 = 2$ for the coefficients
$(-2) + (-1) = -3$ for the exponents

Finally,

$(2 \times 10^{-2}) \times (1 \times 10^{-1}) = 2 \times 10^{-3}$

or

$0.02 \times 0.1 = 0.002$

Note that the coefficient 2 in 2×10^{-3} is still positive. Only the exponent -3 is negative, indicating the fraction 1/1000 for 10^{-3}. However, the value of the complete number 2×10^{-3} is still positive. It is just a positive fraction, less than 1 because of the negative exponent.

Problems 5-N

Multiply in powers of 10.

1. $(2 \times 10^{-2}) \times (3 \times 10^{-4}) =$
2. $(4 \times 10^{-1}) \times (2 \times 10^{-1}) =$
3. $(7 \times 10^{-7}) \times (1 \times 10^{-1}) =$
4. $(3 \times 10^{-7}) \times (2 \times 10^{-8}) =$
5. $(5 \times 10^{-5}) \times (1 \times 10^{-2}) =$
6. $(3 \times 10^{-3}) \times (2 \times 10^{-2}) =$
7. $(7 \times 10^{-2}) \times (1 \times 10^{-2}) =$
8. $(2.5 \times 10^{-4}) \times (2 \times 10^{-4}) =$

For the case of adding positive and negative exponents, take the difference between the two and give it the sign of the larger exponent.

Examples. Multiply the following:

$(4 \times 10^5) \times (2 \times 10^{-3}) = 8 \times 10^2$
$(4 \times 10^{-5}) \times (2 \times 10^3) = 8 \times 10^{-2}$

When there are more than two factors, just keep adding the exponents and multiplying the coefficients.

Example. Multiply the following:

$(2 \times 10^5) \times (1 \times 10^{-4}) \times (3 \times 10^3)$

Answer. $2 \times 1 \times 3 = 6$ for the coefficients
$5 - 4 + 3 = 4$ for the exponents

The final result is

$(2 \times 10^5) \times (1 \times 10^{-4}) \times (3 \times 10^3) = 6 \times 10^4$

Problems 5-O

Do the following multiplications in powers of 10.

1. $(4 \times 10^4) \times (2 \times 10^2) =$
2. $(5 \times 10^3) \times (3 \times 10^3) =$
3. $(3 \times 10^{-2}) \times (2 \times 10^{-1}) =$
4. $(7 \times 10^{-1}) \times (4 \times 10^{-2}) =$
5. $(4 \times 10^5) \times (2 \times 10^{-3}) =$
6. $(2 \times 10^{-5}) \times (3 \times 10^3) =$
7. $(3 \times 10^7) \times (2 \times 10^2) \times (2 \times 10^{-3}) =$
8. $4,000,000 \times 2,000,000 =$
9. $0.001 \times 0.003 =$

10. $4{,}000{,}000 \times 0.000\,002 =$
11. $(7 \times 10^{4}) \times 200 =$
12. $(4 \times 10^{-12}) \times (3 \times 10^{6}) =$

5-6
DIVISION WITH POWERS OF 10

To divide numbers involving powers of 10, divide the coefficients but subtract the exponents.

Example. Divide 6,000,000 by 3000.

Answer. $6{,}000{,}000 = 6 \times 10^{6}$
$3000 = 3 \times 10^{3}$

Then the problem is

$(6 \times 10^{6}) \div (3 \times 10^{3})$
$6 \div 3 = 2$ for the coefficients
$6 - 3 = 3$ for the exponents

The final result is

$(6 \times 10^{6}) \div (3 \times 10^{3}) = 2 \times 10^{3}$

or

$6{,}000{,}000 \div 3000 = 2000$

Only the powers of 10 are subtracted. The coefficients are still divided.

Remember that you must subtract the exponent for the divisor from the exponent for the dividend. If you do the reverse, the result is the reciprocal of the correct answer.

Problems 5-P

Divide in powers of 10.

1. $(8 \times 10^{8}) \div (2 \times 10^{2}) =$
2. $(9 \times 10^{9}) \div (3 \times 10^{3}) =$
3. $(7 \times 10^{18}) \div (2 \times 10^{15}) =$
4. $(6 \times 10^{6}) \div (4 \times 10^{4}) =$
5. $(5 \times 10^{5}) \div (2 \times 10^{2}) =$
6. $(8 \times 10^{8}) \div (2 \times 10^{2}) =$
7. $(7 \times 10^{7}) \div (1 \times 10^{5}) =$
8. $(6 \times 10^{12}) \div (3 \times 10^{8}) =$

When the divisor has a larger exponent, the subtraction results in a negative number for the new exponent.

Example. Divide 8×10^{4} by 2×10^{6}.

Answer. $(8 \times 10^{4}) \div (2 \times 10^{6}) = 4 \times 10^{-2}$

For the coefficients, we have

$8 \div 2 = 4$

For the exponents, we have

$4 - 6 = -2$

The answer of 4×10^{-2} is equal to 0.04.

When the divisor has a negative exponent, which must be subtracted, just change its sign and add.

Example. Divide 6×10^{5} by 2×10^{-3}.

Answer. $(6 \times 10^{5}) \div (2 \times 10^{-3}) = 3 \times 10^{8}$

For the coefficients, we have

$6 \div 2 = 3$

For the exponents, we have

$5 - (-3) = 5 + 3 = 8$

The reason why the answer of 3×10^{8} is larger than the original dividend of 6×10^{5} is that the divisor is a decimal fraction less than 1.

For another possibility of division with negative powers of 10, both exponents can be negative.

Example. Divide 6×10^{-5} by 2×10^{-3}.

Answer. $(6 \times 10^{-5}) \div (2 \times 10^{-3}) = 3 \times 10^{-2}$

For the coefficients, we have

$6 \div 2 = 3$

For the exponents, we have

$-5 - (-3) = -5 + 3 = -2$

Problems 5-Q

Divide, using powers of 10.

1. $(1 \times 10^{8}) \div (1 \times 10^{6}) =$
2. $(1 \times 10^{6}) \div (1 \times 10^{8}) =$
3. $(1 \times 10^{6}) \div (1 \times 10^{-8}) =$
4. $(1 \times 10^{-6}) \div (1 \times 10^{8}) =$
5. $(1 \times 10^{-6}) \div (1 \times 10^{-8}) =$
6. $60{,}000 \div 200 =$
7. $60{,}000 \div 0.002 =$
8. $0.000\,016 \div 8 \times 10^{3} =$

Earlier in the chapter the value of 10^0 was given as 1. The reason for this can be shown by a division problem, remembering that any number divided by itself is equal to 1, or unity.

Example. Using the rules of exponents, divide 1×10^3 by itself.

Answer. $(1 \times 10^{3}) \div (1 \times 10^{3}) = 1 \times 10^{0}$

For the coefficients,

$1 \div 1 = 1$

For the exponents,

$3 - 3 = 0$

Since any number divided by itself must be equal to 1, we can say that

$$\frac{1 \times 10^3}{1 \times 10^3} = 1 \times 10^0 = 1$$

or 10^0 must be 1, as $1 \times 1 = 1$.

In fact, any number taken to the zero power must be 1.

Examples.

$$5^0 = 1$$
$$8^0 = 1$$
$$\left(\frac{1}{2}\right)^0 = 1$$
$$(0.003)^0 = 1$$

5-7 RECIPROCALS WITH POWERS OF 10

Just changing the sign of the exponent converts the base number to its reciprocal.

Example. Find the reciprocal of 10^2.

Answer. The reciprocal can be written as $1/10^2 = 10^{-2}$.

The values of $1/10^2$ and 10^{-2} are both equal to 1/100, as the reciprocal of 100.

Example. Find the reciprocal of 10^{-2}.

Answer. The reciprocal can be written as $1/10^{-2} = 10^2$.

Both values equal 100, as the reciprocal of $^1/_{100}$.

Taking the reciprocal of a number is really a special case of division with the numerator of 1. Remember that 1 can be stated as 10^0. For the example of $1/10^2$, the division is

$$\frac{1}{10^2} = \frac{10^0}{10^2} = 10^{(0-2)} = 10^{-2}$$

In other words, with exponents the reciprocal means a change of sign, because the exponent is subtracted from a zero exponent. This rule applies to any base with any exponent. As another example, the reciprocal of 5^3 or $1/5^3$ is equal to 5^{-3}.

In these examples, the base actually has the coefficient of 1, but the reciprocal is still 1. With any other coefficient, you must find its reciprocal value besides changing the sign of the exponent for the reciprocal of the power of 10.

Example. Find the reciprocal of 2×10^3.

Answer. The reciprocal can be written as $1/(2 \times 10^3)$.

Another way of writing this is $\frac{1}{2} \times 1/10^3 = 0.5 \times 10^{-3}$.

The coefficient is 0.5, as the reciprocal of 2 and 10^{-3} is the reciprocal of 10^3.

Problems 5-R

Give the following reciprocal values without the fraction bar, using powers of 10:

1. $\dfrac{1}{10^{-4}} =$
2. $\dfrac{1}{10{,}000} =$
3. $\dfrac{1}{0.5 \times 10^{-2}} =$
4. $\dfrac{1}{0.25 \times 10^{5}} =$
5. $\dfrac{1}{100} =$
6. $\dfrac{1}{2 \times 10^{2}} =$
7. $\dfrac{1}{10^{7}} =$
8. $\dfrac{1}{0.125 \times 10^{-8}} =$

5-8 ADDITION AND SUBTRACTION WITH POWERS OF 10

Only numbers having the same power of 10 can be added or subtracted directly. Add or subtract only the coefficients but keep the same exponent. The reason for not changing the exponent is to leave the decimal place unchanged.

Example. Add 6×10^3 and 2×10^3.

Answer. $6 + 2 = 8$ for the coefficients

The power of 10 is not changed. Therefore,

$$(6 \times 10^3) + (2 \times 10^3) = 8 \times 10^3$$

Example. Subtract 2×10^3 from 6×10^3.

Answer. $6 - 2 = 4$ for the coefficients

The power of 10 is not changed. Therefore,

$$(6 \times 10^3) - (2 \times 10^3) = 4 \times 10^3$$

If the numbers do not have the same exponent, they must be changed so that the exponents are the same before you can add or subtract. Any exponent can be used, but they must all be the same.

Example. Add 3×10^3 and 4×10^4.

Answer. One of the exponents must be converted to equal the other before the addition can be done.

To change 10^4 to 10^3, you can divide the base by 10, but multiply the coefficient by 10 to keep the value of the number the same. For the base,

$$10^4 \div 10 = 10^3$$

For the coefficient,

$$4 \times 10 = 40$$

Then

$$4 \times 10^4 = 40 \times 10^3$$

Now both of the numbers in the example have the same exponent, and the addition can be carried through.

$$(3 \times 10^3) + (40 \times 10^3) = 43 \times 10^3$$

or,

$$3000 + 40{,}000 = 43{,}000$$

It should be noted that using the same power of 10 corresponds to lining up the decimal points for addition or subtraction.

Problems 5-S

Add or subtract using powers of 10.

1. $(6 \times 10^3) + (2 \times 10^3) =$
2. $(6 \times 10^3) - (2 \times 10^3) =$
3. $(6 \times 10^{-3}) + (2 \times 10^{-3}) =$
4. $0.006 + 0.002 =$
5. $(6 \times 10^3) + (2 \times 10^4) =$
6. $6000 + 20{,}000 =$
7. $(4 \times 10^{-2}) + (5 \times 10^{-3}) =$
8. $0.04 + 0.005 =$

5-9 RAISING AN EXPONENT OF 10 TO A HIGHER POWER

To find the power of a number written with an exponent, raise the coefficient to the indicated power, but just multiply the exponents.

Example. Raise 2×10^2 to the third power.

Answer. This can be written as

$$(2 \times 10^2)^3$$

Raise the coefficient to the power indicated:

$$2^3 = 2 \times 2 \times 2 = 8$$

For the exponents,

$$2 \times 3 = 6$$

Therefore,

$$(2 \times 10^2)^3 = 8 \times 10^6$$

Remember that the coefficient 2 is a factor of the base 10. We can factor out the coefficient separately as

$$(2 \times 10^2)^3 = (2)^3 \times (10^2)^3$$
$$= 8 \times 10^6$$

Problems 5-T

Raise to the power shown.

1. $(10^2)^3 =$
2. $(10^3)^2 =$
3. $(1 \times 10^6)^2 =$
4. $(1 \times 10^2)^7 =$
5. $(10^{-2})^3 =$
6. $(7 \times 10^3)^2 =$
7. $(1 \times 10^7)^2 =$
8. $(4 \times 10^{-6})^2 =$

5-10 TAKING A ROOT WITH POWERS OF 10

To find the root of a number written as a power of 10, take the root of the coefficient but just divide the exponent by the indicated root. This procedure is just the opposite of raising to a higher power.

Example. Find the third root of 8×10^6

Answer. This can be written as

$$\sqrt[3]{8 \times 10^6} = \sqrt[3]{8} \times \sqrt[3]{10^6}$$

The third root of 8, also called the cube root, is found first:

$$\sqrt[3]{8} = 2 \text{ (since } 2 \times 2 \times 2 = 8)$$

For the exponents,

$$6 \div 3 = 2$$

Therefore,

$$\sqrt[3]{8 \times 10^6} = 2 \times 10^2$$

A root can also be expressed as a fractional power, using the reciprocal of the index number. The cube root is the same as the $^1/_3$ power. The square root is the $^1/_2$ power. For instance, $\sqrt{16}$ is the same as $(16)^{1/2}$, which equals 4. Also, $\sqrt[3]{8}$ or $8^{1/3}$ is equal to 2. A root as a fractional power follows the rules of exponents for multiplication and division.

A practical rule with roots is to have the exponent exactly divisible by the index number of the root. In other words, take square roots with an even-numbered exponent and cube roots with an exponent that is a multiple of 3.

Otherwise, taking the root results in a fractional exponent for base 10. For instance, $\sqrt{10^3} = 10^{3/2} = 10^{1.5}$, which has a fractional exponent. Such a number is not an exact multiple of tens. Its value can be determined, though, by logarithms or by using an electronic calculator. Usually, however, the calculations are easier with whole numbers for the exponents. You can always convert the exponent to the desired value before taking the root.

Example. Find the square root of

40×10^5

Answer. This can be written as

$\sqrt{40 \times 10^5}$

Since the solution of this problem will involve dividing the exponent by 2, it would be better if the exponent were an even number rather than 5. The first step then is to convert 10^5 to either 10^4 or 10^6. To convert 10^5 to 10^6, merely multiply 10^5 by 10 and divide the coefficient by 10 so that the original number is not changed:

$$40 \times 10^5 \times \frac{10}{10} = \frac{40}{10} \times 10^5 \times 10$$
$$= 4 \times 10^6$$

Now the exponent is an even number, and the square root can be determined using the rule previously explained.

$$\sqrt{4 \times 10^6} = \sqrt{4} \times \sqrt{10^6}$$
$$= 2 \times 10^3$$

Problems 5-U

Find the indicated root.

1. $\sqrt{4 \times 10^8} =$
2. $\sqrt[3]{27 \times 10^9} =$
3. $(10^{12})^{1/6} =$
4. $\sqrt{1 \times 10^{10}} =$
5. $\sqrt[3]{1 \times 10^9} =$
6. $\sqrt{9 \times 10^{-6}} =$
7. $\sqrt[5]{1 \times 10^{20}} =$
8. $(3.6 \times 10^7)^{1/2} =$

5-11 SUMMARY OF ARITHMETIC OPERATIONS WITH DIGITS 1 AND 0

Working with 1 and 0 often can be confusing, because these digits have special properties.

1. For 1 as a factor:
   (*a*) $1 \times 1 = 1$. Multiplied any number of times, the answer is still 1.
   (*b*) A number multiplied by 1 remains the same. For instance, $10 \times 1 = 10$.
   (*c*) A number divided by 1 remains the same. For instance, $10 \div 1 = 10$.
2. For 0 as a factor:
   (*a*) The product of any number multiplied by 0 is 0.
   (*b*) Division by 0 is not a valid operation. In other words, in arithmetic, we cannot divide by 0.
3. For 1 as a base:
   (*a*) Any power or root of 1 is still 1. For instance, $1^3 = 1$, or $\sqrt{1} = 1$.
4. For 1 as an exponent:
   (*a*) Any base with exponent 1 has the same value as without the exponent. For instance, 10^1 is the same as 10.
5. For 0 as an exponent:
   (*a*) Any base with exponent 0 has the numerical value of 1. The reason is that the exponent 0 results from a base and exponent divided by the same base and exponent. Remember: dividing any number by itself results in the answer 1.

Problems 5-V

Do the indicated operations.

1. $10 \times 0 =$
2. $10 \times 1 =$
3. $(10)^0 \times 5 =$
4. $\sqrt[7]{1} =$
5. $14 \times 0 =$
6. $(14)^0 =$
7. $(1)^7 =$
8. $\dfrac{10^0 \times 1}{1} =$

5-12
COMBINED OPERATIONS

The following problems combine multiplication, division, and addition for powers of 10. Do one operation at a time and use the result for the next operation. It usually is helpful to determine the powers of 10 separately from the calculations for the coefficients.

Note that parentheses can be used instead of the multiplication sign. Numbers not separated by the + or − sign are factors to be multiplied.

Problems 5-W

Answers should be in scientific, or engineering, notation.

1. $(2 \times 10^6)(3 \times 10^3) \div (4 \times 10^4) =$
2. $(2 \times 10^6)(3 \times 10^{-3}) \div (4 \times 10^4) =$
3. $(2 \times 10^6)(3 \times 10^3) \div (4 \times 10^{-4}) =$
4. $(\sqrt{10^8})(10^3) =$
5. $\sqrt{10^4 \times 10^6} =$
6. $(25\sqrt{10^8}) + (3 \times 10^4) =$
7. $\dfrac{10^7}{\sqrt{10^7 \times 10^3}} =$
8. $\dfrac{1}{\sqrt{10^{-3} \times 10^{-5}}} =$
9. $(3 \times 10^3)(2 \times 10^2) + (3 \times 10^5) =$
10. $\dfrac{(4 \times 10^3)(2 \times 10^4)(3 \times 10^5)}{(8 \times 10^8)(1 \times 10^4)} =$

5-13
ELECTRONICS APPLICATIONS

Do the following problems, using powers of 10 for the metric prefixes. The units stated in each problem provide the required unit in the answer.

Problems 5-X

Fill in the missing values.

1. 22 mA × 4 kΩ = ________ V
2. 18 kV ÷ 2 MΩ = ________ A
3. 4 mA × 3 kV = ________ W
4. $(6 \text{ mA})^2 \times 2 \text{ m}\Omega$ = ________ W
5. $\frac{1}{2}$ μs = ________ s
6. $\frac{1}{4}$ MHz = ________ Hz
7. 20 kV ÷ 4 mA = ________ Ω
8. 200 μA = ________ mA
9. 0.022 A × 4000 Ω = ________ V
10. $\dfrac{0.16}{\sqrt{40 \times 10^{-3}\text{ H} \times 4 \times 10^{-12}\text{ F}}} =$ ________ Hz

Chapter 6: Algebra

Algebra is a branch of mathematics that deals with the ordinary arithmetic calculations covered in Chap. 1 but using letter symbols in addition to numbers. The purpose is to make general statements about the values in the form of an equation. For instance, by means of the formula $V = IR$, we can find the voltage V for any values of I and R. Remember that the letters represent not decimal places but entire quantities. The symbol V can have the value of 7, 70, or 700, as examples. Also, letters written together indicate not decimal places but multiplication. IR means I times R. It is the same as $I \times R$ or $I \cdot R$ or $(I)(R)$.

The topics in this chapter are:

6-1 Literal Numbers
6-2 Adding or Subtracting Literal Numbers
6-3 Powers and Roots of Literal Numbers
6-4 Multiplying or Dividing Literal Numbers with Exponents
6-5 Fractions with Literal Numbers
6-6 Terms and Factors
6-7 Polynomials

6-1 LITERAL NUMBERS

Letters such as a, b, c, x, y, and z are used as literal numbers in algebra to represent the decimal values of numerical counting. For instance, a can represent 3, and $2a$ then is 6. The purpose of literal numbers is to state the relationship between quantities in a general form. Letters at the beginning of the alphabet such as a, b, and c usually represent constant or known values, while x, y, and z are for unknown values. Either small letters or capitals can be used.

An example of a general statement in the unknown quantity x can be given as $2x + 3x = 10$. This solution is $5x = 10$ or $x = 2$.

For formulas in electricity and electronics, the letters V, I, R, and P are common. These are used as follows:

I is for the intensity of current. The unit quantity is the ampere, abbreviated A.

V is the voltage that makes current flow. The unit quantity is the volt, abbreviated V.

R is the resistance or opposition to current. The unit quantity is the ohm, represented by the Greek capital letter omega (Ω).

P is for power. The unit is the watt, abbreviated W.

6-2 ADDING OR SUBTRACTING LITERAL NUMBERS

Only the same letters can be added or subtracted.

Examples.

$$5a + 2a = 7a$$
$$5a - 2a = 3a$$
$$5a + 2a + 2b = 7a + 2b$$

(Unlike terms cannot be added.)

Identical letters with different subscripts indicate different literal numbers. To give an example, V_1 and V_2 represent two different quantities.

Identical letters with different exponents also indicate different literal numbers. As an example, a and a^2 are two different literal numbers, which therefore cannot be combined by addition or subtraction.

Problems 6-A

Add or subtract the following literal numbers:

1.	$x + x =$	6.	$2a + 3c + 3a =$
2.	$5a + 3a =$	7.	$6y - y =$
3.	$5a - 3a =$	8.	$V_1 + 5V_1 =$
4.	$6y + y =$	9.	$R_1 + 4R_3 + 2R_1 =$
5.	$5y + 4y =$	10.	$I_1 + 6I_1 =$

6-3 POWERS AND ROOTS OF LITERAL NUMBERS

An exponent of a literal number has the same use as with decimal numbers. For instance, a^2 is equal to $a \times a$. Also, a^3 is $a \times a \times a$. Here the literal number a is the base raised to a higher power.

When the letter has a numerical coefficient included with the literal base, both are raised to the higher power.

Example Square $3a$.

Answer. This is written as $(3a)^2$. Each factor is squared, as

$$3^2 \times a^2 = 9a^2$$

Or the problem can be stated as

$$3a \times 3a = 9a^2$$

If the number were written as $3a^2$, then it would be understood that only the a was squared. Suppose $a = 4$. Then

$$3a^2 = 3 \times 4^2 = 3 \times 16 = 48$$

For $(3a)^2$, however, the value is $(12)^2$, which equals 144.

When an exponent of a literal number is raised to any power, the two exponents are multiplied.

Example. Cube the value of (a^2).

Answer. This is written $(a^2)^3$.

Since the exponents are multiplied, $2 \times 3 = 6$. Therefore,

$$(a^2)^3 = a^6$$

As with decimal numbers, any literal number to the zero power is 1. For example, $x^0 = 1$. Also, $(a^2)^0 = 1$.

Any literal number to the 1 power is the literal number itself. For example, $a^1 = a$ and $(xy)^1 = xy$.

Problems 6-B

Raise to the indicated power.

1.	$(2b)^3 =$	5.	$(3c)^2 =$
2.	$(4y)^2 =$	6.	$(2z)^4 =$
3.	$(5x)^3 =$	7.	$(-3a)^2 =$
4.	$(a^5)^3 =$	8.	$(a^0)^3 =$

A root can also be found for literal numbers. The procedure is the same as with decimal numbers.

Example. Find the fourth root of a^8.

Answer. This is written as $\sqrt[4]{a^8}$. Divide the exponent by the root: $8/4 = 2$. Therefore,

$$\sqrt[4]{a^8} = a^2$$

It should be noted that $\sqrt{a^2}$ is really $\pm a$, as either a or $-a$ squared results in a^2. The problem usually will provide a clue as to whether the answer is + or − or both.

Problems 6-C

Find the indicated root.

1. $\sqrt[3]{8b^3} =$
2. $\sqrt{16y^2} =$
3. $\sqrt{a^6} =$
4. $(a^6)^{1/3} =$
5. $\sqrt{25y^4} =$
6. $\sqrt{a^{14}} =$
7. $\sqrt{10a^2} =$
8. $\sqrt[3]{-8a^3} =$

Problems 6-D

Assuming a is 2 and b is 3, evaluate:

1. $(2b)^2 =$
2. $2b^2 =$
3. $4a^2 + (2a)^2 =$
4. $\sqrt[3]{8b^3} =$
5. $(8b^3)^{1/3} =$
6. $\sqrt{16a^6} =$

6-4 MULTIPLYING OR DIVIDING LITERAL NUMBERS WITH EXPONENTS

Multiply the numerical coefficients of the literal base to obtain the new numerical coefficient, but add the exponents to obtain the new exponent of the base. However, the exponents can be combined this way only if the base is the same for both numbers.

Example. Multiply $6a^2$ by $3a^3$.

Answer. The problem is written $6a^2 \times 3a^3$. Multiply the coefficients:

$6 \times 3 = 18$

Since the base of these exponents is the same for both terms, the exponents are added:

$2 + 3 = 5$

The final answer is, then,

$6a^2 \times 3a^3 = 18a^5$

When no exponent is shown for the base, it actually is 1. For instance, a is the same as a^1. Also, the numerical coefficient is 1 if none is shown, as a^3 is the same as $1a^3$.

Problems 6-E

Multiply the following:

1. $y \times y =$
2. $2y \times 3y =$
3. $4a \times 2a^2 =$
4. $4a^2 \times 2a =$
5. $2a^5 \times 3a^4 =$
6. $3x^2 \times 2x^5 =$
7. $3b \times 5b^3 =$
8. $2I \times 4I =$
9. $3a \times 2b =$
10. $3x^a \times 2x^b =$

To divide literal numbers, divide the numerical coefficients of the literal base to find the new coefficient, and subtract the exponent of the divisor from the exponent of the dividend to obtain the new exponent. Again, the exponents can be combined only for the same base.

Example. Divide $8a^3$ by $2a$.

Answer. The problem is written as

$$8a^3 \div 2a = \frac{8a^3}{2a}$$

The divisor is $2a$. Divide the coefficients: $8 \div 2 = 4$. Then subtract the exponent of the divisor (1 in this case) from the exponent of the dividend.

$3 - 1 = 2$

Therefore,

$$\frac{8a^3}{2a} = 4a^2$$

Problems 6-F

Divide the following:

1. $8a^6 \div 4a^4 =$
2. $12a^3 \div 4a =$
3. $7x^7 \div 2x^2 =$
4. $12b^6 \div 6b^2 =$
5. $8a \div 2b =$
6. $8a^4 \div 4a^7 =$
7. $3y \div 3y =$
8. $3x^a \div 2x^b =$

6-5 FRACTIONS WITH LITERAL NUMBERS

Literal numbers represent entire quantities that could be parts of fractions. Thus, a literal number can be in the numerator or the denominator, for example, $a/2$, $2x/3$, x/y, $(a + b)/c$. Remember also that the *value* of the literal number may in fact be a fraction itself.

Fractions containing literal numbers can be added, subtracted, multiplied, and divided just as can decimal numbers. They can also be raised to a power and have roots extracted.

To add or subtract fractions, the denominators must be the same, whether literal numbers or regular numbers. Once the denominators are the same, the numerators are combined as in ordinary addition and subtraction.

Example. Add $2/a$ and $3/a$.

Answer. $\frac{2}{a} + \frac{3}{a} = \frac{2 + 3}{a} = \frac{5}{a}$

To multiply fractions containing literal numbers, simply multiply the numerators together to obtain the new numerator and multiply the denominators together to obtain the new denominator.

Example. Multiply $2/a \times 3/a$.

Answer. $\frac{2}{a} \times \frac{3}{a} = \frac{6}{a^2}$

To divide fractions, invert the divisor and multiply.

Example. Divide $2/a$ by $3/a$.

Answer. $\frac{2}{a} \div \frac{3}{a} = \frac{2}{a} \times \frac{a}{3} = \frac{2a}{3a} = \frac{2}{3}$

To raise a fraction to a power, both the numerator and denominator are raised to that power.

Example. Find the cube of $a/2$.

Answer. $\left(\frac{a}{2}\right)^3 = \frac{a^3}{2^3} = \frac{a^3}{8}$

To find the root of a fraction, find the root of both numerator and denominator.

Example. Find the cube root of $a^3/8$.

Answer. $\sqrt[3]{\frac{a^3}{8}} = \frac{\sqrt[3]{a^3}}{\sqrt[3]{8}} = \frac{a}{2}$

Problems 6-G

Do the following problems with combined operations:

1. $\frac{x}{2} + \frac{x}{5} =$
2. $\frac{2a}{3} - \frac{a}{3} =$
3. $\frac{2a + 1}{2} + \frac{2a - 1}{4} =$
4. $\left(\frac{x}{3}\right)\left(\frac{y}{2}\right) =$
5. $\left(\frac{2}{3}\right)\left(\frac{a}{b}\right) =$
6. $\frac{a}{b} \div \frac{a}{b} =$
7. $\frac{a}{2} \times \frac{ab}{4} =$
8. $\frac{xy}{a} \div \frac{a}{xy} =$
9. $\frac{a^2b}{a} \times \frac{cd}{ab} =$
10. $\left(\frac{2}{b}\right)^2 =$
11. $\left(\frac{a}{b}\right)^3 =$
12. $\sqrt{\frac{a^2}{b^2}} =$

13. $\sqrt{4a^2} \times 2a =$

14. $\sqrt{\dfrac{4}{b^2}} =$

6-6
TERMS AND FACTORS

Factors are numbers to be multiplied or divided. Terms are numbers that are added or subtracted. For example, in the expression 3 + 2 the 3 and 2 are terms. In the expression $3a + 2a$ the $3a$ and $2a$ are terms, but 3 and a are factors in the term $3a$. The number 3 is further described as the coefficient of the term.

If we consider the expression (3)(2), the 3 and 2 are *factors* of the product 6. This is not the same as the decimal count of 32.

It is important to realize the fundamental difference between factors and terms, because they have different rules of operation in algebraic equations. For instance, in $x = 2 + 3$ all the numbers are terms without any factors. However, in $V = IR$, the I and R are factors of V. The details of working with such equations are explained in Chap. 7.

Problems 6-H

Pick out the terms in the following expressions:

1. $2x + 3$
2. $5y - 4$
3. $abc + a$
4. $xy^2 + 8$
5. $R_1 + R_2$
6. $C_1 + C_2$

Problems 6-I

Pick out the factors in the following expressions:

1. $2x$
2. $5y$
3. abc
4. xy^2
5. R_1R_2
6. $2\pi fL$

Problems 6-J

Evaluate the following when $x = 4$ and $y = 5$:

1. $xy =$
2. $x + y =$
3. $x^2y =$
4. $x^2 + y =$
5. $y^2x =$
6. $y^2 + x =$

6-7
POLYNOMIALS

A *polynomial* is an algebraic expression using literal numbers with more than one term. When only one term is used, it is a *monomial.* Examples of a monomial are a, $2a$, $3x$, $5y^2$, and $4ab$.

A *binomial* has two terms. Examples are $(a + b)$ and $(3x + 5y^2)$. Binomials with the same terms but opposite signs are called *conjugates.* For instance, $(a + b)$ and $(a - b)$ are conjugate binomials.

A *trinomial* has three terms. An example is $a^2 + 2ab + b^2$. The trinomials and binomials are also polynomials. A polynomial can also have more than three terms.

Example. Identify each of the following as either a monomial, binomial, or trinomial:

a; $a + b$; $a^2 + b^2$; ab; $a^3 + bc$; $a + b + c$; $a^2 + b^2 + c^2$; $2a^2 + 3b^2 + 4ab$; $3a^2b^2c^3$

Answer.
a is a monomial (one term: a).
$a + b$ is a binomial (two terms: a and b).
$a^2 + b^2$ is a binomial (two terms: a^2 and b^2).
ab is a monomial (one term: ab).
$a^3 + bc$ is a binomial (two terms: a^3 and bc).
$a + b + c$ is a trinomial (three terms: a, b, and c).
$a^2 + b^2 + c^2$ is a trinomial (three terms: a^2, b^2, and c^2).
$2a^2 + 3b^2 + 4ab$ is a trinomial (three terms: $2a^2$, $3b^2$, and $4ab$).
$3a^2b^2c^3$ is a monomial (one term: $3a^2b^2c^3$).

For multiplication of a polynomial by a monomial, multiply each term in the polynomial by the monomial.

Example. Multiply $x + y$ by $2xy$.

Answer. The problem can be written as

$2xy(x + y)$

Both terms in the binomial $(x + y)$ are multiplied by $2xy$ to produce a new binomial:

$(2xy)(x) + (2xy)(y)$

or $2x^2y + 2xy^2$

Therefore,

$2xy(x + y) = 2x^2y + 2xy^2$

For division of a polynomial by a monomial, divide each term in the polynomial by the monomial.

Example. Divide $x^2 + 2x$ by $3x$.

Answer. The problem can be written as

$$\frac{x^2 + 2x}{3x} = \frac{x^2}{3x} + \frac{2x}{3x}$$
$$= \frac{x}{3} + \frac{2}{3}$$

Combine the terms having a common denominator:

$$\frac{x}{3} + \frac{2}{3} = \frac{x + 2}{3}$$

For multiplication of two polynomials, multiply each term in one polynomial by each term in the other polynomial and add the partial products.

Example. Multiply $x^2 + x + 3$ by $x - 1$.

Answer. The problem can be written similar to the way a numerical multiplication problem is done:

$$\begin{array}{l} x^2 + x + 3 \\ \underline{x \;\; - 1} \end{array}$$

Now proceed in a fashion similar to multiplication in arithmetic but start with the left-hand term. The following multiplications are done and the partial products are recorded as shown:

$$\begin{array}{ll} x \times x^2 = x^3 & -1 \times x^2 = -x^2 \\ x \times x = x^2 & -1 \times x = -x \\ x \times 3 = 3x & -1 \times 3 = -3 \end{array}$$

Multiplying the two polynomials now:

$$\begin{array}{l} x^2 + x + 3 \\ \underline{x \;\; - 1} \\ x^3 + x^2 + 3x \\ \underline{\quad - x^2 - \;\; x - 3} \\ x^3 + 0 + 2x - 3 = x^3 + 2x - 3 \end{array}$$

The methods of division of polynomials are similar to those performed in arithmetic, but they are not shown here because they seldom apply to electronics problems. More details can be found in algebra textbooks.

Problems 6-K

Do the operations indicated.

1. $a^2 \times 2ab =$
2. $a^2 \times (2ab + 4) =$
3. $a^2 \times (2ab - 4) =$
4. $(2a^3b + 4a^2) \div a^2 =$
5. $4abc \div ab =$
6. $(2y + 1) \times y =$
7. $(x + y)^2 =$
8. $(x + y)(x - y) =$

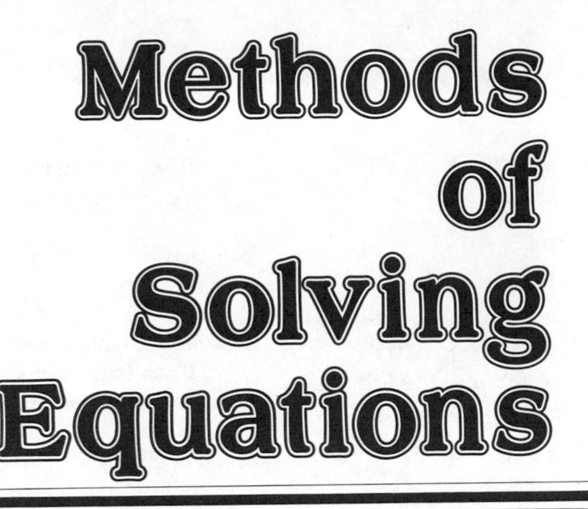

An equation represents a balance of quantities. That is, the combined terms on one side of an equal sign must be equal to the combined terms on the other side of the equal sign. The quantities on either or both sides of the equal sign may contain quantities whose value is not known.

Example. $x = 8 - 5$

The unknown quantity to the left of the equal sign is given a letter symbol. (By custom the unknown quantity is given the letter x or y or z, but actually any letter can be used.) We know $8 - 5$ is 3, and since both sides of the equation must be equal, then

$x = 3$

What we have just done is *solve the equation;* we have found the value of the unknown quantity x.

The topics in this chapter are:

7-1 Operations on Both Sides of an Equation
7-2 Transposing Terms
7-3 Solution of Numerical Equations
7-4 Inverting Factors in Literal Equations

7-1 OPERATIONS ON BOTH SIDES OF AN EQUATION

By rearranging the quantities on both sides of the equal sign, we can solve an equation. In general, we can perform almost any arithmetic operation on one side of the equation as long as we perform the same operation on the other side. The only forbidden operation is dividing by zero.

Addition and Subtraction. Any positive or negative quantity can be added to both sides of an equation without changing the equality.

Example. In the equation $x = 2$, we can add 3 to both x and 2:

$x + 3 = 2 + 3$

The original equation says x is 2. The next equation says that $x + 3$, or 5, is equal to $2 + 3$, or 5. For both equations, x is 2 and the left side equals the right side. Thus, the addition of 3 to both sides of the equation had no effect on the value of x.

Problems 7-A

Add -2 to both sides in the following equations:

1.	$x + 2 = 5$	4.	$a = b$
2.	$y + 2 = 9$	5.	$I + 2 = 6$
3.	$x^2 + 2 = 18$	6.	$c + 2 = -6$

Reversing the Sides. Since the equality applies both ways, the left and right sides of an equation can be interchanged.

Example. $x = 2 + 3$ is the same equality as $2 + 3 = x$
$x + 5 = 2x + 2$ is the same as $2x + 2 = x + 5$

Multiplication. Any number can be used to multiply both sides of an equation. However, every term in the equation must be multiplied by this number. If only some terms are multiplied and others are not, the original equation will be changed and the value of the unknown quantity will be affected.

Example. $y = 3$

If we multiply both sides by 5, we get

$y \times 5 = 3 \times 5$

Since y is 3 from the original equation, we find 3×5 on the left and 3×5 on the right. Both sides equal 15.

If we had multiplied only the left side of the equation, we would have gotten

$y \times 5 = 3$

But since y is 3, the new "equation" would be $3 \times 5 = 3$, which of course is false.

If we had multiplied only the right side of the equation, we would have gotten

$y = 3 \times 5$

or $y = 15$

But y was 3 in our original equation, so we have changed the value of y.

Problems 7-B

Multiply both sides by 4 in the following equations:

1.	$\frac{x}{4} = 1$	4.	$\frac{x}{4} = 2 + 3$
2.	$\frac{y}{4} = 2$	5.	$\frac{v}{4} = \frac{2}{4}$
3.	$\frac{x}{4} = 0$	6.	$0.25y = 3$

Division. Just as we can multiply every quantity in an equation by a number without changing the equation, so too can we divide every quantity by a number (except zero) and leave the equation unchanged.

Example. Divide both sides of the following equation by 3:

$3x = 6$

Answer. Dividing by 3,

$$\frac{3x}{3} = \frac{6}{3}$$
$$x = 2$$

Note that this division actually led to the solution of the equation. By dividing we found the value of x was 2.

Problems 7-C
Divide both sides by 4 in the following equations:

1. $4x = 4$
2. $4y = 16$
3. $4V = 12$
4. $4I = 9$
5. $4x = 8a$
6. $4x = 8 + 4$

Reciprocals. You can take the reciprocal of both sides of an equation. However, the entire side must be inverted, not just a part.

Example. Find the value of x in the following equation:

$$\frac{1}{x} = \frac{1}{2+3}$$

Answer. If we take the reciprocal of both sides of the equation, we obtain

$$x = 2 + 3$$

or $x = 5$

Note that this is not the same as

$$\frac{1}{x} = \frac{1}{2} + \frac{1}{3}$$

The reason is that the fraction bar is a sign of group for 2 + 3.

Problems 7-D
Take the reciprocals of both sides in the following equations:

1. $\frac{1}{x} = \frac{1}{5}$
2. $\frac{1}{y} = \frac{1}{4}$
3. $\frac{1}{x} = \frac{1}{2a}$
4. $\frac{1}{x} = \frac{1}{4+5}$
5. $\frac{1}{a} = \frac{1}{10^{-6}}$
6. $\frac{1}{x} = 5 \times 10^5$

Powers and Roots. Finding powers and roots is often useful when solving equations. You can take any power or root provided the same operation is performed on both sides of the equation.

Example. Solve $x^2 = 9 + 7$.

Answer. First combine terms:

$$x^2 = 16$$

Take the square root of both sides of the equation:

$$\sqrt{x^2} = \sqrt{16}$$
$$x = 4$$

Note, however, that $(-4) \times (-4) = 16$, and so -4 is also a solution of the equation. Square roots always have both plus and minus roots. The physical conditions of the problem will usually establish whether the positive or negative root is appropriate or whether they are both valid.

Problems 7-E
Take the square root of both sides in the following equations:

1. $x^2 = 16$
2. $y^2 = 9$
3. $x^2 = 4b^2$
4. $x^2 = (2 + 3)^2$
5. $I^2 = 19 + 6$
6. $a^2 = b^2$

7-2 TRANSPOSING TERMS

Transposing means moving a term from one side of the equation to the other. When a term is transposed, its sign must be changed, either from + to − or from − to +.

The transposing of terms is often necessary for the numerical solution of an equation.

Example. Solve the equation

$$x + 4 = 10$$

Answer. If the $+4$ is transposed to the right side of the equation, its sign is changed and the new equation is

$x = 10 - 4$
$x = 6$

The reason why the sign must be changed in transposing terms is that the equality must be maintained. In this example, transposing 4 is the same as adding -4 to both sides. On the left side $+4$ and -4 cancel each other, leaving only the x term.

Problems 7-F

Transpose all the literal terms to the left side and all the numbers to the right side in the following equations:

1. $x - 2 = 4$
2. $y^2 - 4 = 5$
3. $3x + 7 = 2x + 8$
4. $2y + 2 = y - 3$
5. $2a - 5 = b + 3$
6. $x + y^2 = y^2 + 7$
7. $V + 3 = 5$
8. $I - 6 = 2$

7-3 SOLUTION OF NUMERICAL EQUATIONS

The first operation in solving equations is to group all unknowns on one side of the equal sign and all numbers on the other. Then combine terms for one numerical value and one term for the unknown. Finally, divide both sides of the equation by the numerical coefficient of the unknown.

Example. Solve the equation

$6x - 8 = 4x + 2$

Answer. Transpose the unknown terms to the left side and all numbers to the right side. Remember to change the sign of a transposed term.

$6x - 4x = 8 + 2$

Combine terms:

$2x = 10$

Divide by 2 (the coefficient of x):

$$\frac{2x}{2} = \frac{10}{2}$$

$x = 5$

Check the answer by substituting 5 in the original equation:

$$\begin{aligned} 6x - 8 &= 4x + 2 \\ 6(5) - 8 &= 4(5) + 2 \\ 30 - 8 &= 20 + 2 \\ 22 &= 22 \end{aligned}$$

The fact that each side has the same value shows that the solution of $x = 5$ satisfies the equality.

There are several additional points to keep in mind for solving the next group of equations. When the equation has signs of grouping, they must be removed before combining terms. The reason is that only terms can be transposed, not factors. The terms 0 and 1 can be transposed the same way as other numbers. Finally, a numerical solution can have a negative value.

Problems 7-G

Solve the following equations for x:

1. $6x = 4x + 8$
2. $4a - 32 = 2a$
3. $6x - 3 = 2x + 21$
4. $8x - 6 = 12x - 14$
5. $5x - 3 - 12 - 8x = 0$
6. $14 + 5 + 6 = 3x + 2x$
7. $5(x - 2) = 30$

8. $5 - (x - 2) = 30$
9. $4(x - 1) + 8 = 6$
10. $(2x + 4)(2x - 4) = 4x^2 + 2x$

The method of transposing terms can also be used for solving literal equations that have two or more terms on one side. This idea is illustrated in the next set of problems.

Example. Solve the equation $a + b = c - a$ for the value of a.

Answer. In this equation all the a terms are transposed to the left side of the equation and the non-a terms to the right side.

$$a + a = c - b$$
$$2a = c - b$$

Dividing both sides by the coefficient of a:

$$\frac{2a}{2} = \frac{c - b}{2}$$

$$a = \frac{c - b}{2}$$

Problems 7-H

Solve the following equations:

1. $c = a + b$ (Solve for a.)
2. $R_T = R_1 + R_2$ (Solve for R_1.)
3. $V_T = V_1 + V_2$ (Solve for V_1.)
4. $R^2 = Z^2 - X^2$ (Solve for Z^2.)
5. $Z^2 = R^2 + X^2$ (Solve for Z.)
6. $C_T = C_1 + C_2$ (Solve for C_2.)

7-4 INVERTING FACTORS IN LITERAL EQUATIONS

Many electronics formulas are in the form of a literal equation that has only one term on each side, but the term itself may have factors. An important example is $V = IR$ for Ohm's law. The general form of such a formula can be stated as the equation $a = bx$. In order to solve for x alone, its factor b must be moved to the other side of the equation. However, b here is a factor, not a term. The rules for moving a factor are as follows:

1. The factor becomes inverted between numerator and denominator on opposite sides of the equation.
2. The sign of the inverted factor is *not changed.*

When there is no denominator shown, it actually is 1. Equation $a = bx$ can be written with denominators as

$$\frac{a}{1} = \frac{bx}{1}$$

In this example, the factor b in the numerator at the right becomes inverted as a factor of 1 in the denominator at the left. The solution for x then is $a/b = x$, or $x = a/b$. This operation is equivalent to dividing both sides by b in the original equation. For this reason, the sign is not changed when inverting a factor from one side to the other, as there is no transposing of terms.

The rule for inverting factors means that the numerator on one side of the equation and the denominator on the other side can be cross-multiplied.

Example. Show the equation $a/b = c/d$ in another form with two factors on each side of the equation. Then solve for a, b, c, and d individually.

Answer. The equation $a/b = c/d$ when cross-multiplied becomes

$$ad = cb$$

Solving for a,

$$a = \frac{cb}{d}$$

Solving for d,

$$d = \frac{cb}{a}$$

Solving for b,

$$\frac{ad}{c} = b \quad \text{or} \quad b = \frac{ad}{c}$$

Solving for c,

$$\frac{ad}{b} = c \quad \text{or} \quad c = \frac{ad}{b}$$

In summary, a factor can be moved *on the diagonal,* from numerator on one side to denominator on the other side, or the opposite way, without a change in sign.

Problems 7-I

Solve the following equations:

1. $ab = c$ (Solve for a.)
2. $a + b = c$ (Solve for a.)
3. $V = IR$ (Solve for I.)
4. $I = \frac{V}{R}$ (Solve for V.)
5. $P = VI$ (Solve for I.)
6. $I = \frac{P}{V}$ (Solve for P.)
7. $Q = CV$ (Solve for C.)
8. $C = \frac{Q}{V}$ (Solve for Q.)
9. $P = I^2R$ (Solve for I.)
10. $l = \pi d$ (Solve for π.)
11. $y = abx$ (Solve for x.)
12. $X = 2\pi fL$ (Solve for L.)
13. $y = \frac{a - b}{x}$ (Solve for x.)
14. $4x^2 = 100$ (Solve for x.)

Trigonometry

Trigonometry is the study of the properties of a triangle. A triangle is made up of three straight lines, called the *sides* of the triangle. Where two lines meet, they enclose an angle. A triangle thus contains three angles.

The topics in this chapter are:

8-1 Types of Angles
8-2 The Right Triangle
8-3 Trigonometric Functions
8-4 Table of Trigonometric Functions
8-5 Angles of More Than 90°
8-6 Radian Measure of Angles

8-1 TYPES OF ANGLES

An angle is formed when two lines meet. In Fig. 8-1, when the line OP_1 is hinged at the origin O and rotates to the position P_2, the line generates the angle P_1OP_2. The Greek letter symbols θ (theta) and ϕ (phi) are generally used with the angle sign $\angle$. In this example P_1OP_2 is indicated as $\theta = 60°$.

The unit of 1°, or one degree, is $^1/_{360}$ of the rotation around a circle. The complete circle, therefore, consists of 360°. In this example, θ is 60°, because the amount of rotation is $^1/_6$ of a complete circle, equal to 360°/6, or 60°.

By convention, counterclockwise rotation, as in Fig. 8-1, is the positive direction for measuring angles. This rotation corresponds to motion upward or to the right for the positive direction for linear measurements.

Example. How many degrees are there in $^1/_8$ rotation of a circle?

Answer. Since the complete circle contains 360°,

$^1/_8 \times 360° = 45°$

A negative angle can be generated by clockwise rotation. In Fig. 8-1, if the line OP_1 were rotated the same amount but in the opposite direction, the angle would be −60°.

Problems 8-A

How many degrees are there in each of the following rotations?

1. $^1/_4$ circle
2. $^1/_2$ circle
3. $^3/_4$ circle
4. $^1/_{10}$ circle
5. $^1/_3$ circle
6. $^1/_6$ circle
7. $^1/_5$ circle
8. Full circle

The angle of 90° is called a *right angle.* One side is upright compared to the other, as shown

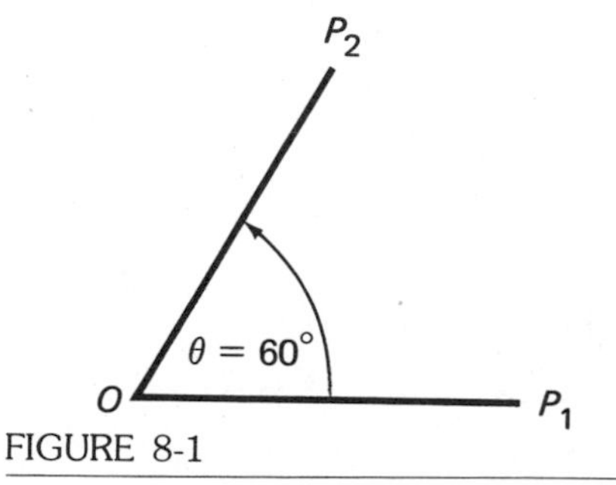

FIGURE 8-1

Generating the angle θ, positive in the counterclockwise direction.

in Fig. 8-2. Since 90° is one-quarter of a circle, the perpendicular sides are in *quadrature*.

Angles of less than 90° are acute. For instance, 60° is an acute angle. An angle of more than 90° is *obtuse*. An example is 120°.

Problems 8-B

State whether each of the following is an acute, obtuse, or right angle:

1.	17°	3.	90°	5.	45°
2.	64°	4.	150°	6.	100°

Two angles whose sum is equal to 90° are *complementary*. For instance, 30° and 60° are complementary angles; one angle is the complement of the other. Two angles whose sum is equal to 180° are *supplementary*. Examples of supplementary angles are 120° and 60°.

Problems 8-C

Give the complement of the following angles:

1.	30°	3.	45°	5.	20°
2.	60°	4.	17°	6.	53°

An angle of less than 1° can be divided into decimal parts. For instance, 0.5° is one-half a degree. Also, 26.5° is midway between 26° and 27°. This method of describing an angle in decimal fractions is called *decitrig* notation.

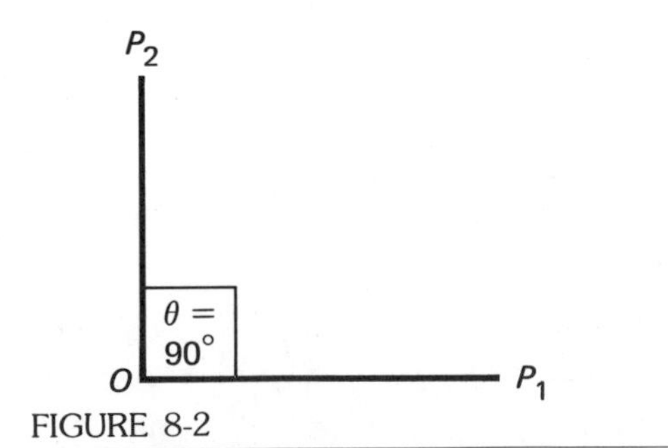

FIGURE 8-2

Angle θ of 90° is a right angle with perpendicular sides.

Angles may also be divided into minutes and seconds. There are 60 minutes in a degree and 60 seconds in a minute. This method of dividing an angle into 60 subdivisions is called the *sexagesimal* notation.

Example. Write 26°30′0″ using the decitrig notation.

Answer. Since 30′ = ½° = 0.5°,

26°30′0″ = 26.5°

The decitrig form of notation, such as 26.5°, is used for slide rules and electronic calculators. Calculations with angles are much simpler in decitrig form.

To add or subtract angles just combine their numerical values.

Example. Add 26.5° and 32.7°.

Answer. Align the decimal points and add as usual.

$$\begin{array}{r} 26.5^\circ \\ +32.7^\circ \\ \hline 59.2^\circ \end{array}$$

Problems 8-D

Add or subtract the following angles, as indicated:

1.	30° + 10° =	5.	70° + 50° =
2.	12.5° + 8° =	6.	120° − 70° =
3.	84° − 14° =	7.	2° + 3° =
4.	14.5° + 2°30′ =	8.	24° − 30° =

8-2 THE RIGHT TRIANGLE

In the analysis of ac circuits, the calculations make extensive use of trigonometry with right triangles. The reason is that the perpendicular sides of the triangle can represent currents or voltages 90° out of phase. Refer to Fig. 8-3. The altitude a is perpendicular to the base b. These

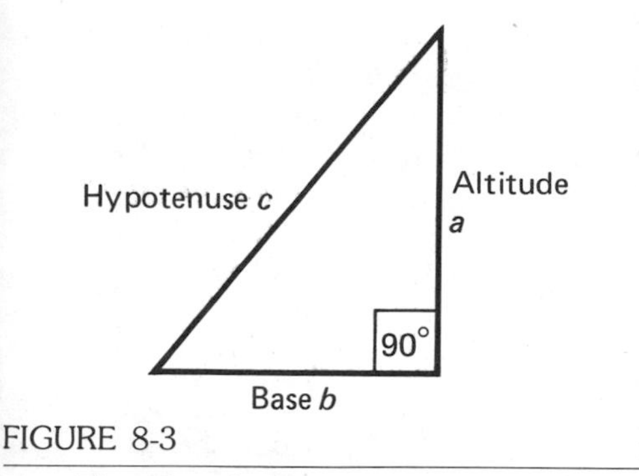

FIGURE 8-3

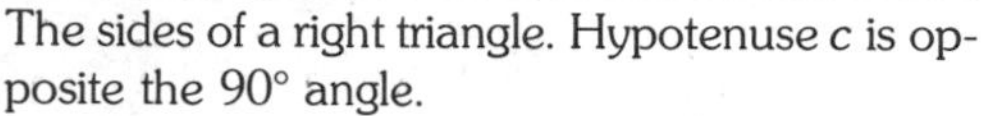

The sides of a right triangle. Hypotenuse c is opposite the 90° angle.

two sides form a right angle. The side opposite the 90° angle is called the *hypotenuse,* indicated by the letter c.

The Pythagorean theorem from ancient Greek geometry states that the square of the hypotenuse is equal to the sum of the squares of the other two sides. This equality is illustrated in Fig. 8-4, with squares for all the sides of the triangle. As a formula,

$$c^2 = a^2 + b^2 \tag{8-1}$$

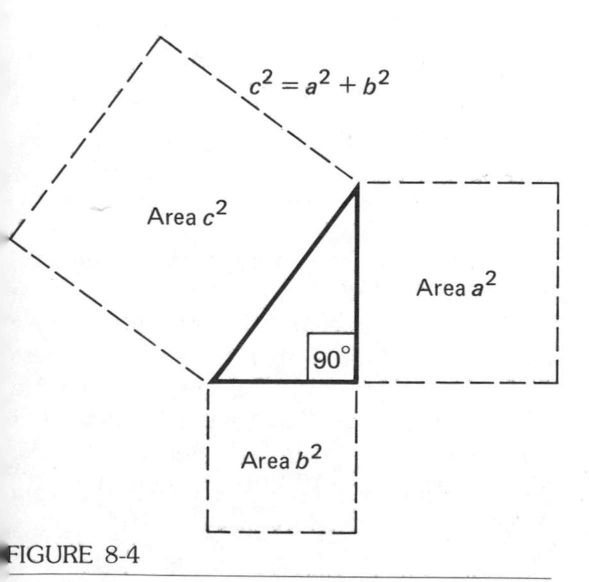

FIGURE 8-4

Illustrating the Pythagorean theorem for a right triangle: $c^2 = a^2 + b^2$.

In order to find the length of the hypotenuse, it is necessary to find the square root of c^2. Taking the square roots of both sides of the equation, we get

$$c = \sqrt{a^2 + b^2} \tag{8-2}$$

To summarize, the length of the hypotenuse is calculated as follows:

1. Square each of the other two sides.
2. Add the squares.
3. Take the square root of the sum of the squares.

Example. In Fig. 8-5*a*, one side of the right triangle is 4 units long while the other side is 3 units long. Find the hypotenuse.

Answer. Using the formula

$$c = \sqrt{a^2 + b^2}$$

square each side:

$$4^2 = 16$$
$$3^2 = 9$$
$$c = \sqrt{16 + 9} = \sqrt{25}$$

Find the square root of the sum of the squares:

$$c = \sqrt{25} = 5 \text{ units}$$

This type of triangle is called a 3:4:5 right triangle. When the perpendicular sides are in the ratio 3:4, as in Fig. 8-5*a*, or 4:3, as in Fig. 8-5*b*, the hypotenuse must be in the ratio of 5.

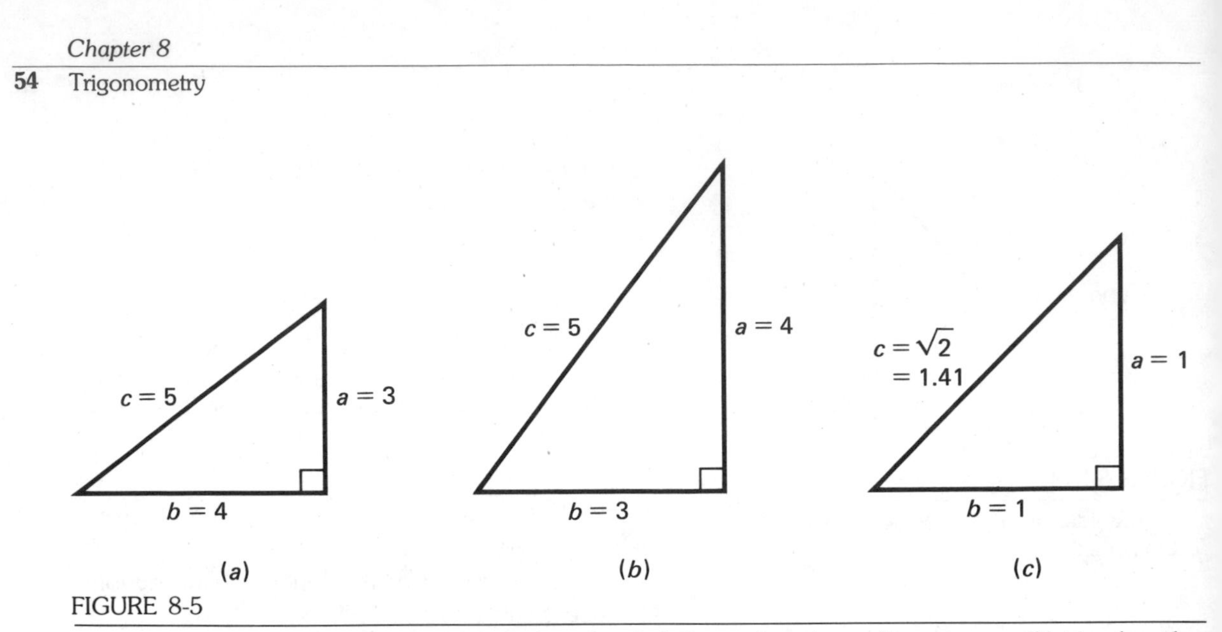

FIGURE 8-5

Special types of right triangle: (*a*) triangle with a 3:4:5 ratio for sides and hypotenuse; (*b*) triangle with a 4:3:5 ratio; (*c*) equilateral triangle with equal sides.

Example. Assume that two sides of a right triangle are 15 and 20. Find the hypotenuse.

Answer.

$$c = \sqrt{a^2 + b^2} = \sqrt{15^2 + 20^2}$$
$$c = \sqrt{225 + 400} = \sqrt{625}$$
$$c = 25$$

The three sides of the triangle are thus, 15, 20, and 25. If we divided each of these sides by 5,

$$\frac{15}{5} = 3 \qquad \frac{20}{5} = 4 \qquad \frac{25}{5} = 5$$

we would end up with the basic 3:4:5 right triangle. In other words, the actual lengths may vary, but their ratio will remain the same in this type of triangle.

Another special type of right triangle is the equilateral triangle, shown in Fig. 8-5*c*, with two equal sides. Here the hypotenuse must be $\sqrt{2}$, or 1.41, times greater than either side.

Example. Two sides of an equilateral right triangle are equal to 10. Find the hypotenuse.

Answer. The formula $c = \sqrt{a^2 + b^2}$ will, of course, lead to the answer. However, the hypotenuse must be 1.41 times the length of one side. Thus

$$c = 1.41 \times 10 = 14.1$$

As a check the formula will be used:

$$c = \sqrt{10^2 + 10^2} = \sqrt{100 + 100}$$
$$c = \sqrt{200} = \sqrt{2 \times 100} = \sqrt{2} \times \sqrt{100}$$
$$c = 1.41 \times 10 = 14.1$$

This checks with the original calculation.

Formulas 8-1 and 8-2 apply to any right triangle. To check the answer to a right-triangle problem, it should be noted that the hypotenuse must be longer than either of the sides but less than their sum. For instance, in a 3:4:5 triangle the hypotenuse 5 is greater than either side, 3 or 4, but less than their sum of 7. Also, in Fig. 8-5*c*, 14.1 is more than 10 but less than 10 + 10, or 20.

Problems 8-E

Find the hypotenuse c of the following right triangles:

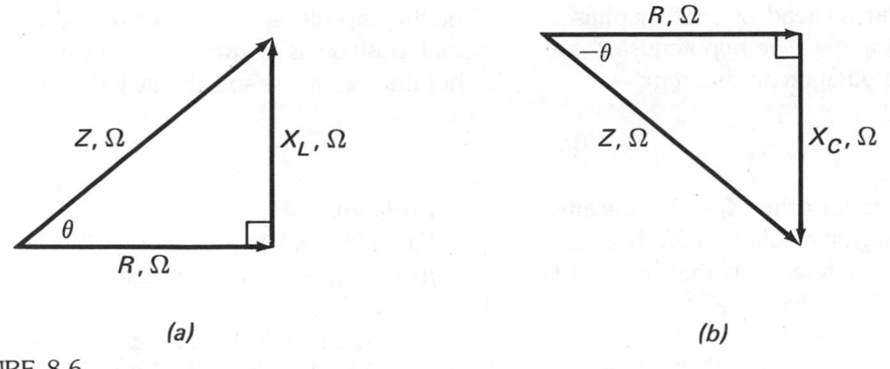

FIGURE 8-6

Using hypotenuse of right triangle for electrical impedance Z. (a) Sides represent phasors for inductive reactance X_L at 90° with respect to resistance R. (b) Capacitive reactance X_C at −90°.

1. $a = 6, b = 8$
2. $a = 8, b = 6$
3. $a = 5, b = 5$
4. $a = 2, b = 4$
5. $a = 4, b = 2$
6. $a = 10, b = 1$
7. $a = 1, b = 10$
8. $a = 10, b = 10$
9. $a = 4, b = 5$
10. $a = 2, b = 8$

Formulas 8-1 and 8-2 can be transposed to find one side when the other side and the hypotenuse are given. Then,

$$a = \sqrt{c^2 - b^2} \tag{8-3}$$

$$b = \sqrt{c^2 - a^2} \tag{8-4}$$

Note that in the radical, a^2 or b^2 must be subtracted from c^2, because the hypotenuse c is the longest side of the triangle.

Example. Find the length of the other side of a right triangle in which the hypotenuse is 13 and one side is 12.

Answer.

$$\begin{aligned} a &= \sqrt{c^2 - b^2} \\ a &= \sqrt{13^2 - 12^2} \\ &= \sqrt{169 - 144} \\ &= \sqrt{25} \\ a &= 5 \end{aligned}$$

Problems 8-F

Find the missing side a or b in the following triangles:

1. $c = 14.14$, $a = 10$
2. $c = 14.14$, $b = 10$
3. $c = 5$, $b = 4$
4. $c = 5$, $a = 3$
5. $c = 4.47$, $a = 2$
6. $c = 4.47$, $b = 2$
7. $c = 6.4$, $b = 5$
8. $c = 8.25$, $a = 2$

The way that the right triangle is used for ac analysis is illustrated in Fig. 8-6 for a series circuit. The perpendicular sides represent reactance X and resistance R which are 90° out of phase. Then the hypotenuse is their phasor sum, which is the resultant impedance Z. All these quantities are in units of ohms (Ω).

In Fig. 8-6a, the perpendicular vector is the inductive reactance X_L of a coil. This right triangle is shown upward with an angle of 90°, for the phase between X_L and R. In Fig. 8-6b, the triangle for X_C is shown downward with an angle of −90°, because an X_C voltage lags an R voltage. The X_C is the reactance of a capacitor.

Actually both triangles are diagrams for the addition of phasors by starting the tail of one phasor from the arrowhead of another phasor, with the specified angle. The hypotenuse Z can be found by the Pythagorean theorem:

$$Z = \sqrt{R^2 + X^2} \tag{8-5}$$

This formula applies for either X_L or X_C, because the square of a negative value for X_C becomes positive. The only difference is that angle θ is positive for X_L but negative for X_C.

Problems 8-G

Find the magnitude of Z, without the phase angle, for the following values of X and R:

1. $X_L = 6, R = 8$
2. $X_C = 6, R = 8$
3. $X_L = 5, R = 5$
4. $X_C = 5, R = 5$
5. $X_L = 2, R = 3.75$
6. $X_C = 4, R = 5$
7. $X_L = 1, R = 10$
8. $X_L = 10, R = 1$

An important characteristic of any triangle is that the sum of the three angles must equal 180°. For a right triangle, one angle is 90°. Therefore, the other two angles must total 90°, which makes them complementary. As an example, in Fig. 8-7 with $\theta = 30°$, the complementary angle ϕ is 60°.

It should be noted that θ is in the *standard reference position,* with one side horizontal. Then the opposite side is the altitude a. This horizontal position is assumed to mean angle θ, rather than ϕ, unless specifically indicated otherwise.

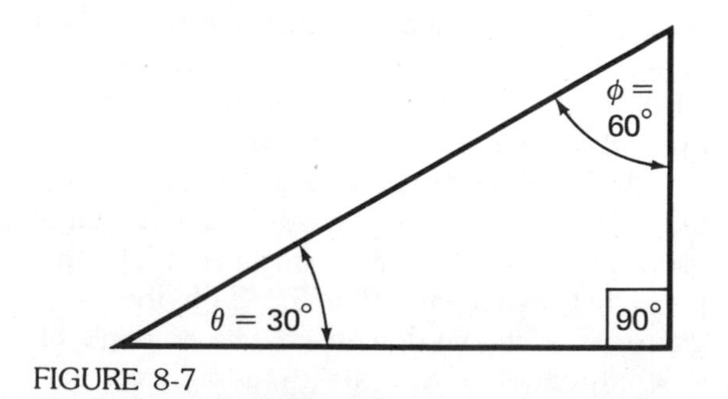

FIGURE 8-7

Right triangle showing complementary angles θ and ϕ; 30° and 60°, respectively, they total 90°. Angle θ is in the standard reference position.

Problems 8-H

Find either angle θ or ϕ, as indicated in the following, for a right triangle:

1. $\theta = 30°, \phi = ?$
2. $\theta = 60°, \phi = ?$
3. $\phi = 45°, \theta = ?$
4. $\phi = 20°, \theta = ?$
5. $\theta = 5°, \phi = ?$
6. $\phi = 5°, \theta = ?$
7. $\theta = 53°, \phi = ?$
8. $\phi = 12.5°, \theta = ?$

8-3 TRIGONOMETRIC FUNCTIONS

A mathematical function is a fixed relation between two quantities. A trigonometric function defines an angle in a triangle in terms of the sides. Referring to Fig. 8-8 a triangle having sides a, b, and hypotenuse c, for angle θ in the standard reference position, the three main trigonometric functions are defined as follows:

Sine of θ, or sin θ,

$$= \frac{\text{opposite side}}{\text{hypotenuse}} = \frac{a}{c} \tag{8-6}$$

Cosine of θ, or cos θ,

$$= \frac{\text{adjacent side}}{\text{hypotenuse}} = \frac{b}{c} \tag{8-7}$$

Tangent of θ, or tan θ,

$$= \frac{\text{opposite side}}{\text{adjacent side}} = \frac{a}{b} \tag{8-8}$$

Side b is the adjacent side for angle θ because b is one side of the angle. Side a is the opposite side because it is not part of the angle. The hypotenuse is always opposite the right angle.

These formulas can be applied to the 3:4:5 triangle in Fig. 8-8. For the trigonometric functions of angle θ,

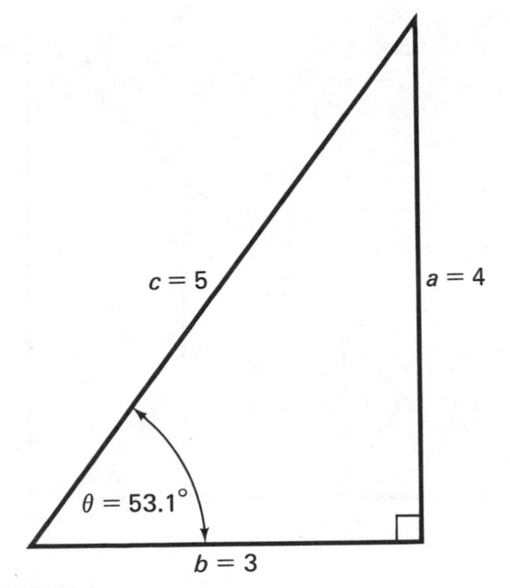

FIGURE 8-8
Right triangle for defining sin θ, cos θ, and tan θ.

$$\sin\theta = \frac{\text{opposite side}}{\text{hypotenuse}} = \frac{4}{5} = 0.8$$

$$\cos\theta = \frac{\text{adjacent side}}{\text{hypotenuse}} = \frac{3}{5} = 0.6$$

$$\tan\theta = \frac{\text{opposite side}}{\text{adjacent side}} = \frac{4}{3} = 1.33$$

Note that the functions are numerical values for the ratio of two sides, not degrees like angle θ. However, the value of the function specifies the angle. This example is for $\theta = 53.1°$, as shown in Fig. 8-8.

The angle of 53.1°, as one example, has these values for sine, cosine, and tangent in any triangle. Any angle can be specified either in degrees or in terms of its trigonometric functions.

Problems 8-I

Find sin θ, cos θ, or tan θ, as indicated, in the following right triangles with the given opposite side a, adjacent side b, and hypotenuse c.

1. $a = 2, c = 2; \sin\theta = ?$
2. $b = 1, c = 2; \cos\theta = ?$
3. $a = 4, c = 5; \sin\theta = ?$
4. $b = 4, c = 5; \cos\theta = ?$
5. $a = 6, b = 8, c = 10; \sin\theta = ?$
6. $a = 6, b = 8, c = 10; \tan\theta = ?$
7. $a = 2, b = 2, c = 2.8; \tan\theta = ?$
8. $a = 10, b = 10, c = 14.14; \tan\theta = ?$

It should be noted that there are three more trigonometric functions, the reciprocals of sin θ, cos θ, and tan θ. These reciprocal functions are:

$$\text{Cotangent of } \theta = \cot\theta = \frac{\text{adjacent side}}{\text{opposite side}} = \frac{b}{a} = \frac{1}{\tan\theta}$$

$$\text{Secant of } \theta = \sec\theta = \frac{\text{hypotenuse}}{\text{adjacent side}} = \frac{c}{b} = \frac{1}{\cos\theta}$$

$$\text{Cosecant of } \theta = \csc \theta = \frac{\text{hypotenuse}}{\text{opposite side}} = \frac{c}{a}$$
$$= \frac{1}{\sin \theta}$$

Reciprocal functions invert the ratio of the two sides. For the tangent and cotangent, as an example, if $\tan \theta = 3/4$, then $\cot \theta = 4/3$. Or, if $\tan \theta = 0.75$ then $\cot \theta = 1/0.75 = 1.33$.

Problems 8-J

Find the reciprocal functions.

1. $\tan \theta = 2$, $\cot \theta = ?$
2. $\cot \theta = 0.5$, $\tan \theta = ?$
3. $\tan \theta = 1$, $\cot \theta = ?$
4. $\tan \theta = 8$, $\cot \theta = ?$

8-4 TABLE OF TRIGONOMETRIC FUNCTIONS

The ratio of the sides for a specific angle will always be the same for any size triangle. This idea is illustrated in Fig. 8-9. For the tangent ratio of opposite side to adjacent side, the angle of 45° shown has $\tan \theta$ equal to $2/2$, $3/3$, or $4/4$. They all equal 1, which is tan 45°. Similarly, the sine or cosine of 45° will always be 0.707. Therefore, the angle is specified by its function, or the function defines the angle.

The numerical values are listed in a trigonometric table, such as Table 8-1, for angles from 0 to 90°. These values can also be obtained with an electronic calculator, of the scientific type with trigonometric functions, or from a decitrig slide rule.

Table 8-1 lists angles from 0 to 44° going down the page and then starts at the top again for 45 to 90°. Each angle has its sine, cosine, and tangent listed in a horizontal row. Using the angle of 30° as an example in the table, sin 30° = 0.5000, cos 30° = 0.8660, and tan 30° = 0.5774. As another example, for 60°

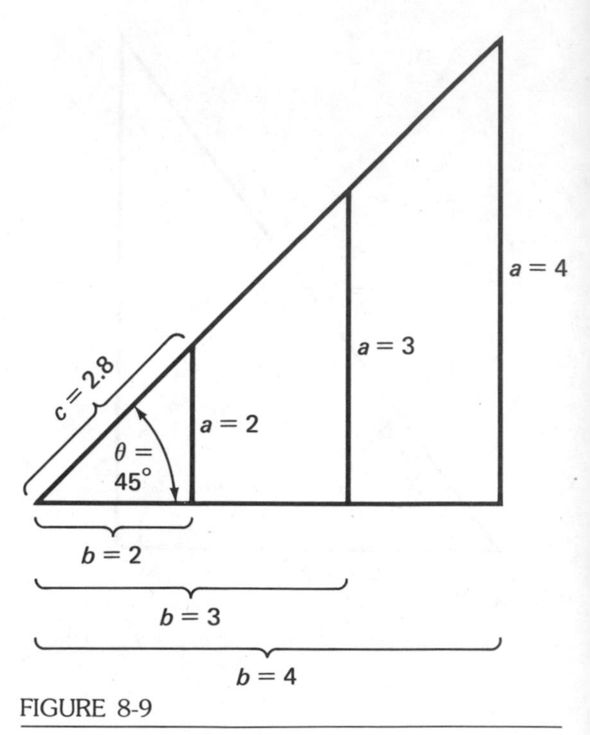

FIGURE 8-9

Here $\theta = 45°$, and $\tan 45° = 2/2, 3/3, 4/4$, or 1, for any size triangle.

in the right-hand side of the table, sin 60° = 0.8660, cos 60° = 0.5000, and tan 60° = 1.7321.

Any one angle has the three functions, but we generally use only one at a time. When the problem involves the opposite and adjacent sides without the hypotenuse, then $\tan \theta$ is useful.

We can determine the functions from the angle or find the angle from the function. For instance, if we know that the angle is 45°, we can determine that tan 45° = 1. Or if we know that $\tan \theta = 1$, then θ must be 45°.

Note that each function is not in degree units but is just a numerical ratio without any units. It is a pure number, because the units for the sides of the triangle cancel in their ratio. The

TABLE 8-1. Trigonometric Functions

ANGLE	SIN	COS	TAN	ANGLE	SIN	COS	TAN
0°	0.0000	1.000	0.0000	45°	0.7071	0.7071	1.0000
1	.0175	.9998	.0175	46	.7193	.6947	1.0355
2	.0349	.9994	.0349	47	.7314	.6820	1.0724
3	.0523	.9986	.0524	48	.7431	.6691	1.1106
4	.0698	.9976	.0699	49	.7547	.6561	1.1504
5	.0872	.9962	.0875	50	.7660	.6428	1.1918
6	.1045	.9945	.1051	51	.7771	.6293	1.2349
7	.1219	.9925	.1228	52	.7880	.6157	1.2799
8	.1392	.9903	.1405	53	.7986	.6018	1.3270
9	.1564	.9877	.1584	54	.8090	.5878	1.3764
10	.1736	.9848	.1763	55	.8192	.5736	1.4281
11	.1908	.9816	.1944	56	.8290	.5592	1.4826
12	.2079	.9781	.2126	57	.8387	.5446	1.5399
13	.2250	.9744	.2309	58	.8480	.5299	1.6003
14	.2419	.9703	.2493	59	.8572	.5150	1.6643
15	.2588	.9659	.2679	60	.8660	.5000	1.7321
16	.2756	.9613	.2867	61	.8746	.4848	1.8040
17	.2924	.9563	.3057	62	.8829	.4695	1.8807
18	.3090	.9511	.3249	63	.8910	.4540	1.9626
19	.3256	.9455	.3443	64	.8988	.4384	2.0503
20	.3420	.9397	.3640	65	.9063	.4226	2.1445
21	.3584	.9336	.3839	66	.9135	.4067	2.2460
22	.3746	.9272	.4040	67	.9205	.3907	2.3559
23	.3907	.9205	.4245	68	.9272	.3746	2.4751
24	.4067	.9135	.4452	69	.9336	.3584	2.6051
25	.4226	.9063	.4663	70	.9397	.3420	2.7475
26	.4384	.8988	.4877	71	.9455	.3256	2.9042
27	.4540	.8910	.5095	72	.9511	.3090	3.0777
28	.4695	.8829	.5317	73	.9563	.2924	3.2709
29	.4848	.8746	.5543	74	.9613	.2756	3.4874
30	.5000	.8660	.5774	75	.9659	.2588	3.7321
31	.5150	.8572	.6009	76	.9703	.2419	4.0108
32	.5299	.8480	.6249	77	.9744	.2250	4.3315
33	.5446	.8387	.6494	78	.9781	.2079	4.7046
34	.5592	.8290	.6745	79	.9816	.1908	5.1446
35	.5736	.8192	.7002	80	.9848	.1736	5.6713
36	.5878	.8090	.7265	81	.9877	.1564	6.3138
37	.6018	.7986	.7536	82	.9903	.1392	7.1154
38	.6157	.7880	.7813	83	.9925	.1219	8.1443
39	.6293	.7771	.8098	84	.9945	.1045	9.5144
40	.6428	.7660	.8391	85	.9962	.0872	11.43
41	.6561	.7547	.8693	86	.9976	.0698	14.30
42	.6691	.7431	.9004	87	.9986	.0523	19.08
43	.6820	.7314	.9325	88	.9994	.0349	28.64
44	.6947	.7193	.9657	89	.9998	.0175	57.29
				90	1.0000	.0000	∞

angle θ itself is in degrees, but its trigonometric functions are not.

Problems 8-K

Find the value of the trigonometric functions.

1.	sin 10° =	7.	tan 45° =
2.	sin 30° =	8.	tan 70° =
3.	cos 60° =	9.	sin 60° =
4.	tan 43° =	10.	cos 30° =
5.	sin 45° =	11.	tan 18° =
6.	cos 45° =	12.	sin 58° =

Another use of the table is to determine the angle when the value of the function is calculated from the known sides of a triangle. For instance, when the ratio of the opposite side to the hypotenuse is 0.5, then sin θ must be 0.5. Reference to the table will show that sin θ is 0.5 when $\theta = 30°$. Thus, to find an angle when two sides are known, look up the function value in the table and find its corresponding angle.

Example. Find the angles whose function values are as follows:

$\sin\theta = 0.5$; then $\theta = 30°$
$\cos\theta = 0.5$; then $\theta = 60°$
$\tan\theta = 1$; then $\theta = 45°$

Notice that the sum of the angles whose sine and cosine functions are equal is 90°. This is always true. Thus for $\sin\theta = 0.5000$ and $\cos\theta = 0.5000$, $\theta = 30°$ in the first case and $\theta = 60°$ in the second; $30° + 60° = 90°$.

Problems 8-L

Find angle θ from the function given.

1.	$\sin\theta = 0.5446$	5.	$\tan\theta = 0.0349$
2.	$\cos\theta = 0.7193$	6.	$\tan\theta = 0.8391$
3.	$\sin\theta = 0.7071$	7.	$\tan\theta = 1$
4.	$\cos\theta = 0.7071$	8.	$\tan\theta = 1.1918$

In Table 8-1, sin θ increases from 0 for 0° to 1 for 90°. For 0°, the opposite side is 0 and the sine ratio is 0. For larger angles, the opposite side becomes longer and the sine ratio increases. However, the maximum sine ratio is 1, because no side of the triangle can be longer than the hypotenuse. For 90°, the opposite side can be considered the same as the hypotenuse for the sine ratio of 1.

Problems 8-M

Find sin θ for the following angles:

1.	0°	5.	60°
2.	20°	6.	70°
3.	30°	7.	80°
4.	45°	8.	90°

The values for cos θ start from 1 as a maximum value for 0°. The reason is that for 0°, the adjacent side can be considered the same as the hypotenuse and the cosine ratio is unity. Then cos θ decreases for larger angles as the adjacent side becomes smaller. At 90°, the adjacent side can be considered zero and $\cos\theta = 0$.

Problems 8-N

Find cos θ for the following angles:

1.	0°	5.	60°
2.	20°	6.	70°
3.	30°	7.	80°
4.	45°	8.	90°

Again, note that equal values of sine and cosine functions produce complementary angles. For instance, sin 30° is the same as cos 60°, both equal to 0.5.

The tangent function is probably used most often, to find the phase angle in ac circuits. The reason is that tan θ involves only the perpendicular sides, which correspond to resistance and reactance.

The values of tan θ increase without limit as angle θ increases toward 90°, because the oppo-

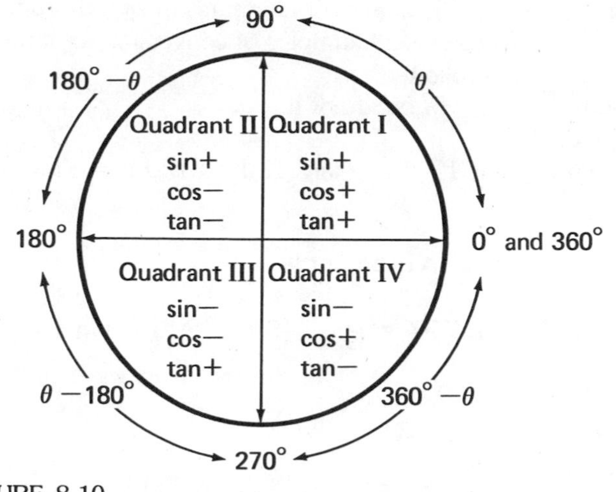

FIGURE 8-10

Angles in the four quadrants of a circle.

site side becomes longer. However, tan θ values should be considered in two parts, below and above 45°:

1. Up to 45°, tan θ increases from 0 to 1. At 45°, the opposite and adjacent sides are equal for the ratio of 1 for tan θ. A subdivision here is tangent values of less than 0.1 for angles less than 5.7°.
2. Above 45°, tan θ is more than 1 because the opposite side is longer than the adjacent side. A subdivision here is tangent values of more than 10 for angles above 84.3°. At 90°, the tangent ratio is infinite (∞), as the opposite side then is infinitely long.

The main thing to remember about the tangent function is that when the opposite side is longer, tan θ is more than 1; when the adjacent side is longer, tan θ is less than 1.

Problems 8-O

Find tan θ for the following angles:

1. 0° 2. 6°
3. 30° 6. 72°
4. 45° 7. 84°
5. 60° 8. 88°

Problems 8-P

Calculate tan θ and find θ from the sides in the following examples. (a = *opposite side,* b = *adjacent side.)*

1. $b = 4, a = 1$ 5. $b = 4, a = 5$
2. $b = 4, a = 2$ 6. $b = 4, a = 6$
3. $b = 4, a = 3$ 7. $b = 4, a = 8$
4. $b = 4, a = 4$ 8. $b = 4, a = 40$

8-5
ANGLES OF MORE THAN 90°

The full circle of 360° is divided into four quadrants of 90° each, as shown in Fig. 8-10. In the counterclockwise direction, the quadrants are as follows:

I 0 to 90°
II 90 to 180°
III 180 to 270°
IV 270 to 360° (or 0°)

After 360° or any multiple of 360°, the angles just repeat the values from 0°.

To use the trigonometric functions for angles in quadrants II, III, and IV, these values can be converted to equivalent angles in quadrant I. The following rules apply for θ greater than 90°:

In quadrant II, use $180° - \theta$
In quadrant III, use $\theta - 180°$
In quadrant IV, use $360° - \theta$

Note that the conversions are with respect to the horizontal axis only, using either 180° or 360° as the reference. This way the obtuse angle is always subtracted from a larger angle.

Example. Convert 135° to an angle of less than 90°.

Answer. The angle is in quadrant II. Therefore, using the formula,

$$\begin{aligned} \text{Angle} &= 180 - \theta \\ &= 180 - 135 \\ &= 45° \end{aligned}$$

Problems 8-Q

Convert to an acute angle of less than 90°.

1. 160° 5. 210°
2. 200° 6. 330°
3. 240° 7. 315°
4. 150° 8. 390°

It is also necessary to consider the sign or polarity of the function in each quadrant.

All the functions are positive in quadrant I.

The sine is also positive in quadrant II, where the vertical ordinate is still positive. However, the sine is negative in quadrants III and IV, where the vertical axis (ordinate) is negative.

The cosine is negative in quadrants II and III, where the horizontal axis (abscissa) is negative, but positive in quadrants IV and I.

The tangent alternates in sign through the quadrants. Examples of conversion for tan θ are as follows:

In quadrant II,

$$\tan 120° = -\tan (180° - 120°) = -\tan 60° = -1.7321$$

In quadrant III,

$$\tan 240° = \tan (240° - 180°) = \tan 60° = +1.7321$$

In quadrant IV,

$$\tan 300° = -\tan (360° - 300°) = -\tan 60° = -1.7321$$

Example. Find the value of the sine, cosine, and tangent for an angle of 315°, in quadrant IV.

Answer.

$$\begin{aligned} \sin 315° &= -\sin (360° - 315°) \\ &= -\sin 45° = -0.7071 \\ \cos 315° &= +\cos 45° = +0.7071 \\ \tan 315° &= -\tan 45° = -1.000 \end{aligned}$$

Problems 8-R

Find sin θ, cos θ, and tan θ with the correct sign for the following obtuse angles:

1. $\theta = 160°$ 3. $\theta = 210°$
2. $\theta = 200°$ 4. $\theta = 330°$

8-6 RADIAN MEASURE OF ANGLES

In circular measure it is convenient to use a specific unit angle called the *radian* (abbreviated rad). Its convenience is due to the fact that a radian is the angular part of a circle that includes an arc equal to the radius, as shown in Fig. 8-11.

The circumference around a circle is $2\pi r$ in length, where r is the radius. The Greek letter π is the ratio of the circumference to the diameter for any circle. Then π is a constant value at 3.14, and $2\pi = 6.28$.

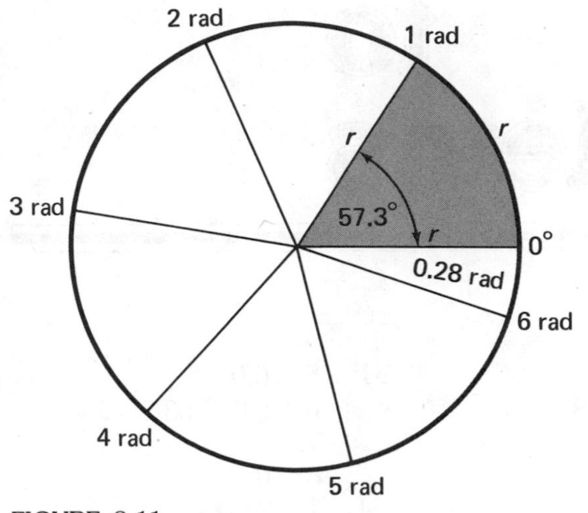

FIGURE 8-11

One radian angle is 57.3°. The complete circle includes 2π rad, or 6.28 rad.

Since the circumference is $2\pi r$ and since one radian uses one length of r, the complete circle includes 2π rad. A circle also is 360°. Therefore, 2π rad = 360°. Or one radian is 360° divided by 2π. This division gives

$$1 \text{ rad} = \frac{360°}{6.28} = 57.3°$$

The quadrants of a circle can be considered conveniently in radian angles instead of degrees. The corresponding values are shown in the table below:

Quadrant	I				II	III	IV
Degrees	0 30	45	60	90	180	270	360

Degrees	0	30	45	60	90	180	270	360
Radians	0	$\frac{\pi}{6}$	$\frac{\pi}{4}$	$\frac{\pi}{3}$	$\frac{\pi}{2}$	π	$\frac{3\pi}{2}$	2π

Quadrant I spans 0–90, II spans 90–180, III spans 180–270, IV spans 270–360.

In general, to convert degrees into radians, divide the angles in degrees by 57.3°/rad.

Example. Express 86° in radians.

Answer. $86° \div 57.3°/\text{rad} = \frac{86° \text{ rad}}{57.3°} = 1.5 \text{ rad}$

Note that the degree units cancel.

To convert radians into degrees, multiply by 57.3°/rad.

Example. Convert 1.5 rad to degrees.

Answer. $1.5 \text{ rad} \times 57.3°/\text{rad} = 1.5 \times 57.3°$
$= 86°$

Note that the radian units cancel.

Problems 8-S

Convert to radians. Write the answer in terms of π.

1. 45° =
2. 90° =
3. 180° =
4. 270° =
5. 360° =
6. 720° =

Problems 8-T

Convert to degrees.

1. $\frac{\pi}{2}$ rad =
2. 1.75 rad =
3. π rad =
4. 3.5 rad =
5. $\frac{3\pi}{2}$ rad =
6. 5.24 rad =
7. 5.76 rad =
8. 2π rad =

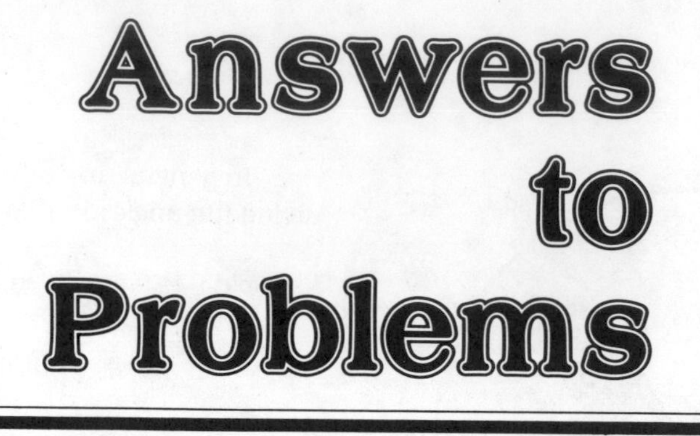

CHAPTER 1

1-A
1. Two
2. Three
3. Four
4. Two
5. Five
6. Seven

1-B
1. 25.0
2. 225.0
3. 205.0
4. 3007.0

1-C
1. 0.6
2. 0.62
3. 0.007
4. 8.3
5. 0.205
6. 0.000 007

1-D
1. 230
2. 1232
3. 889
4. 764
5. 435.7
6. 432.5
7. 796.4
8. 634.34

1-E
1. 2.2
2. 4
3. 1624.83
4. 24
5. 17
6. 566
7. 0.0076
8. 0.08
9. 63
10. 258,444

I-F
1. 9.6
2. 9.6
3. 8.4
4. 27.06
5. 80
6. 80
7. 99.9
8. 924

1-G
1. 32
2. 3.2
3. 0.32
4. 230
5. 3
6. 30
7. 24
8. 0.5
9. 2
10. 0.5

1-H
1. 200
2. 580
3. 3000
4. 5432
5. 5.7
6. 0.036
7. 42
8. 4.2

1-I
1. 2
2. 5.8
3. 3
4. 5.432
5. 0.0057
6. 0.000 036
7. 4.2
8. 0.42

1-J
1. 22,000
2. 12,000
3. 300
4. 0.088
5. 0.48
6. 0.48

1-K
1. 9
2. 16
3. 25
4. 49
5. 3
6. 4
7. 5
8. 7

1-L
1. 10
2. 83.75
3. 0.6
4. 63.7 V (but 74.3 if 0 reading omitted)
5. 314.5 Ω (but 86.0 if 1000 reading omitted)

1-M
1. 5.2
2. 7.98
3. 2.7
4. 31.1

1-N
1. 12

1-N 2. 8
3. 7
4. 0
5. 45
6. 29
7. 12
8. 30

1-O 1. 10 V
2. 10 V
3. 9 V
4. 0.3 V
5. 470 V
6. 0.02 V
7. 12,000 V
8. 14.14 V

1-P 1. 20 W
2. 50 W
3. 1 W
4. 1000 W
5. 20 W
6. 243 W

1-Q 1. 1280
2. 1300
3. 1280
4. 0.128
5. 500
6. 490
7. 480
8. 0.333

CHAPTER 2

2-A 1. 13
2. 3
3. 3
4. −13
5. 7
6. −7
7. 11
8. −8

2-B 1. 3
2. 13
3. −13
4. 3
5. −7
6. 23
7. 23
8. −7

2-C 1. −12
2. −12
3. 9
4. −14
5. −27
6. 20
7. −12
8. −3

2-D 1. −4
2. −4
3. 4
4. −2
5. 4
6. −3
7. −64
8. 144

2-E 1. −24
2. 36
3. 1
4. −8
5. 24
6. 36
7. −27
8. −24

CHAPTER 3

3-A 1. $\frac{2}{9}$
2. $\frac{2}{21}$
3. $\frac{8}{27}$
4. $\frac{9}{28}$
5. $\frac{6}{45}$
6. $\frac{8}{21}$

3-B 1. $\frac{5}{12}$
2. $\frac{9}{14}$
3. $\frac{27}{64}$
4. $\frac{21}{25}$

3-C 1. $\frac{2}{3}$
2. $\frac{1}{2}$
3. $\frac{1}{4}$
4. $\frac{3}{7}$

3-D 1. $\frac{3}{6}$ and $\frac{1}{6}$
2. $\frac{2}{14}$ and $\frac{3}{14}$
3. $\frac{4}{10}$ and $\frac{3}{10}$
4. $\frac{4}{6}$ and $\frac{1}{6}$

3-E 1. $1\frac{1}{6}$
2. $4\frac{2}{3}$
3. $1\frac{1}{5}$
4. $2\frac{1}{4}$

3-F 1. $\frac{1}{2}$
2. 3
3. $\frac{6}{7}$
4. $\frac{1}{9}$
5. $\frac{3}{25}$
6. $\frac{2}{21}$

3-G 1. $\frac{5}{9}$
2. 1
3. $1\frac{1}{3}$
4. $\frac{2}{9}$
5. $\frac{1}{9}$
6. $\frac{5}{9}$

3-H 1. $\frac{7}{9}$
2. $\frac{1}{9}$
3. $\frac{4}{9}$
4. $\frac{7}{20}$
5. 1
6. 4
7. $\frac{11}{49}$
8. 0
9. $\frac{3}{5}$
10. $\frac{6}{7}$

3-I 1. $\frac{6}{9}$ or $\frac{2}{3}$

3-I
2. $-{}^{8}/_{81}$
3. $-{}^{36}/_{18}$ or -2
4. ${}^{8}/_{81}$

3-J
1. 0.2
2. 0.02
3. 0.2
4. 2
5. 0.111
6. 0.159

3-K
1. 0.9
2. 0.4
3. 0.25
4. 0.06
5. -0.06
6. 0.006

CHAPTER 4

4-A
1. 8
2. 9
3. 10,000
4. 125
5. 36
6. 64
7. 27
8. 81

4-B
1. 2
2. 3
3. 10
4. 5
5. 6
6. 4
7. 8
8. 2

4-C
1. 25
2. 9
3. 1
4. 49
5. 100
6. 81

4-D
1. 5
2. 6
3. 9
4. 10
5. 3
6. 2

4-E
1. 9
2. 9
3. 8
4. -1
5. $+1$
6. 1
7. 16
8. -64

4-F
1. 2
2. -2
3. -3
4. 5
5. $j8$
6. -4
7. 5
8. 2

4-G
1. ${}^{4}/_{9}$
2. ${}^{9}/_{49}$
3. ${}^{1}/_{16}$
4. ${}^{1}/_{9}$
5. ${}^{2}/_{3}$
6. ${}^{3}/_{7}$
7. ${}^{1}/_{2}$
8. ${}^{1}/_{3}$

4-H
1. ${}^{1}/_{2}$
2. ${}^{2}/_{3}$
3. ${}^{1}/_{5}$
4. ${}^{3}/_{7}$
5. ${}^{1}/_{10}$
6. 10

4-I
1. 6561
2. 729
3. 7
4. 256
5. 8
6. 100
7. 25
8. 384

4-J
1. 64
2. 6561
3. 50,625
4. 49×10^{8}
5. 8
6. 6561
7. 12
8. 7×10^{4}

4-K
1. 49
2. 49×10^{6}
3. 1
4. 25×10^{-8}
5. 36
6. 361

4-L
1. 5
2. 6
3. 3
4. 3×10^{3}
5. 4
6. 5

4-M
1. 5
2. 5
3. 5.6
4. 5
5. 8.48
6. 8.94

4-N
1. 3
2. 10
3. 6
4. 5.7
5. 6.92
6. 14.4

CHAPTER 5

5-A
1. 10^{2}
2. 10^{3}
3. 10^{4}
4. 10^{5}
5. 10^{6}
6. 10^{7}

5-B
1. 1000
2. 100
3. 1,000,000
4. 10,000

5-C
1. 10^{-3}
2. 10^{-3}
3. 10^{-2}
4. 10^{2}
5. 10^{-6}
6. 10^{-6}

5-D
1. 0.001
2. 0.1
3. 0.01
4. 0.000 001

5-E
1. $^1/_{1000}$
2. $^1/_{10}$
3. $^1/_{100}$
4. $^1/_{1,000,000}$

5-F
1. 10^{4}
2. 10^{-2}
3. 10^{-5}
4. 10^{7}
5. 10^{8}
6. 10^{-6}
7. 10^{-4}
8. 10^{6}

5-G
1. 10^{2}
2. 10^{-4}
3. 10^{-7}
4. 10^{5}
5. 10^{6}
6. 10^{-8}
7. 10^{-6}
8. 10^{4}

5-H See Problems 5-I

5-I See Problems 5-H

5-J
1. 18×10^{3}
2. 42×10^{-3}
3. 8×10^{6}
4. 7.6×10^{3}
5. 2×10^{3}
6. 2×10^{6}
7. 2×10^{-3}
8. 2×10^{-6}

5-K See Problems 5-L

5-L See Problems 5-K

5-M
1. 6×10^{6}
2. 8×10^{2}
3. 7×10^{8}
4. 6×10^{15}
5. 5×10^{7}
6. 6×10^{5}
7. 7×10^{4}
8. 5×10^{8}

5-N
1. 6×10^{-6}
2. 8×10^{-2}
3. 7×10^{-8}
4. 6×10^{-15}
5. 5×10^{-7}
6. 6×10^{-5}
7. 7×10^{-4}
8. 5×10^{-8}

5-O
1. 8×10^{6}
2. 15×10^{6}
3. 6×10^{-3}
4. 28×10^{-3}
5. 8×10^{2}
6. 6×10^{-2}
7. 12×10^{6}
8. 8×10^{12}
9. 3×10^{-6}
10. 8
11. 14×10^{6}
12. 12×10^{-6}

5-P
1. 4×10^{6}
2. 3×10^{6}
3. 3.5×10^{3}
4. 1.5×10^{2}
5. 2.5×10^{3}
6. 4×10^{6}
7. 7×10^{2}
8. 2×10^{4}

5-Q
1. 1×10^{2}
2. 1×10^{-2}
3. 1×10^{14}
4. 1×10^{-14}
5. 1×10^{2}
6. 3×10^{2}

5-Q
7. 3×10^{7}
8. 2×10^{-9}

5-R
1. 10^{4}
2. 10^{-4}
3. 2×10^{2}
4. 4×10^{-5}
5. 10^{-2}
6. 0.5×10^{-2}
7. 1×10^{-7}
8. 8×10^{8}

5-S
1. 8×10^{3}
2. 4×10^{3}
3. 8×10^{-3}
4. 8×10^{-3}
5. 2.6×10^{4}
6. 2.6×10^{4}
7. 4.5×10^{-2}
8. 4.5×10^{-2}

5-T
1. 10^{6}
2. 10^{6}
3. 1×10^{12}
4. 1×10^{14}
5. 10^{-6}
6. 49×10^{6}
7. 1×10^{14}
8. 16×10^{-12}

5-U
1. 2×10^{4}
2. 3×10^{3}
3. 10^{2}
4. 1×10^{5}
5. 1×10^{3}
6. 3×10^{-3}
7. 1×10^{4}
8. 6×10^{3}

5-V
1. 0
2. 10
3. 5
4. 1
5. 0
6. 1
7. 1
8. 1

5-W
1. 1.5×10^{5}
2. 1.5×10^{-1}

5-W 3. 1.5×10^{13}
4. 10^7
5. 10^5
6. 28×10^4
7. 10^2
8. 10^4
9. 9×10^5
10. 3

5-X 1. 88 V
2. 9×10^{-3} A
3. 12 W
4. 72 W
5. 0.5×10^{-6} s
6. 0.25×10^6 Hz
7. 5×10^6 Ω
8. 0.2 mA
9. 88 V
10. 0.4×10^6 Hz

CHAPTER 6

6-A 1. $2x$
2. $8a$
3. $2a$
4. $7y$
5. $9y$
6. $5a + 3c$
7. $5y$
8. $6V_1$
9. $3R_1 + 4R_3$
10. $7I_1$

6-B 1. $8b^3$
2. $16y^2$
3. $125x^3$
4. a^{15}
5. $9c^2$
6. $16z^4$
7. $9a^2$
8. 1

6-C 1. $2b$
2. $4y$
3. a^2
4. a^2
5. $5y^2$
6. a^7
7. $3.16a$
8. $-2a$

6-D 1. 36
2. 18
3. 32
4. 6
5. 6
6. 32

6-E 1. y^2
2. $6y^2$
3. $8a^3$
4. $8a^3$
5. $6a^9$
6. $6x^7$
7. $15b^4$
8. $8I^2$
9. $6ab$
10. $6x^{(a+b)}$

6-F 1. $2a^2$
2. $3a^2$
3. $3.5x^5$
4. $2b^4$
5. $\frac{4a}{b}$
6. $2a^{-3}$
7. 1
8. $1.5x^{(a-b)}$

6-G 1. $\frac{7x}{10}$
2. $\frac{a}{3}$
3. $\frac{6a+1}{4}$
4. $\frac{xy}{6}$
5. $\frac{2a}{3b}$
6. 1
7. $\frac{a^2b}{8}$
8. $\frac{x^2y^2}{a^2} = \left(\frac{xy}{a}\right)^2$
9. cd
10. $\frac{4}{b^2}$
11. $\frac{a^3}{b^3}$
12. $\frac{a}{b}$
13. $4a^2$
14. $\frac{2}{b}$

6-H 1. $2x$ and 3
2. $5y$ and -4
3. abc and a
4. xy^2 and 8
5. R_1 and R_2
6. C_1 and C_2

6-I 1. 2 and x
2. 5 and y
3. a, b, and c
4. x and y^2
5. R_1 and R_2
6. 2, π, f, and L

6-J 1. 20
2. 9
3. 80
4. 21
5. 100
6. 29

6-K 1. $2a^3b$
2. $2a^3b + 4a^2$
3. $2a^3b - 4a^2$
4. $2ab + 4$
5. $4c$
6. $2y^2 + y$
7. $x^2 + 2xy + y^2$
8. $x^2 - y^2$

CHAPTER 7

7-A
1. $x = 3$
2. $y = 7$
3. $x^2 = 16$
4. $a - 2 = b - 2$
5. $I = 4$
6. $c = -8$

7-B
1. $x = 4$
2. $y = 8$
3. $x = 0$
4. $x = 20$
5. $v = 2$
6. $y = 12$

7-C
1. $x = 1$
2. $y = 4$
3. $v = 3$
4. $I = 2.25$
5. $x = 2a$
6. $x = 3$

7-D
1. $x = 5$
2. $y = 4$
3. $x = 2a$
4. $x = 9$
5. $a = 10^6$
6. $x = 0.2 \times 10^{-5}$

7-E
1. $x = 4$
2. $y = 3$
3. $x = 2b$
4. $x = 5$
5. $I = 5$
6. $a = b$

7-F
1. $x = 6$
2. $y^2 = 9$
3. $x = 1$
4. $y = -5$
5. $2a - b = 8$
6. $x = 7$
7. $V = 2$
8. $I = 8$

7-G
1. $x = 4$
2. $a = 16$
3. $x = 6$
4. $x = 2$
5. $x = -5$
6. $5 = x$
7. $x = 8$
8. $x = -23$
9. $x = 0.5$
10. $-8 = x$

7-H
1. $a = c - b$
2. $R_1 = R_T - R_2$
3. $V_1 = V_T - V_2$
4. $Z^2 = R^2 + X^2$
5. $Z = \sqrt{R^2 + X^2}$
6. $C_2 = C_T - C_1$

7-I
1. $a = \frac{c}{b}$
2. $a = c - b$
3. $I = \frac{V}{R}$
4. $V = IR$
5. $I = \frac{P}{V}$
6. $P = IV$
7. $C = \frac{Q}{V}$
8. $Q = CV$
9. $I = \sqrt{\frac{P}{R}}$
10. $\pi = \frac{l}{d}$
11. $x = \frac{y}{ab}$
12. $L = \frac{x}{(2\pi f)}$
13. $x = \frac{(a - b)}{y}$
14. $x = 5$

CHAPTER 8

8-A
1. 90°
2. 180°
3. 270°
4. 36°
5. 120°
6. 60°
7. 72°
8. 360°

8-B
1. Acute
2. Acute
3. Right
4. Obtuse
5. Acute
6. Obtuse

8-C
1. 60°
2. 30°
3. 45°
4. 73°
5. 70°
6. 37°

8-D
1. 40°
2. 20.5°
3. 70°
4. 17°
5. 120°
6. 50°
7. 5°
8. −6°

8-E
1. 10
2. 10
3. 7.07
4. 4.47
5. 4.47
6. 10.05
7. 10.05
8. 14.14
9. 6.4
10. 8.25

8-F
1. $b = 10$
2. $a = 10$
3. $a = 3$
4. $b = 4$

8-F
5. $b = 4$
6. $a = 4$
7. $a = 4$
8. $b = 8$

8-G
1. $Z = 10\ \Omega$
2. $Z = 10\ \Omega$
3. $Z = 7.07\ \Omega$
4. $Z = 7.07\ \Omega$
5. $Z = 4.25\ \Omega$
6. $Z = 6.4\ \Omega$
7. $Z = 10.05\ \Omega$
8. $Z = 10.05\ \Omega$

8-H
1. $\phi = 60°$
2. $\phi = 30°$
3. $\theta = 45°$
4. $\theta = 70°$
5. $\phi = 85°$
6. $\theta = 85°$
7. $\phi = 37°$
8. $\theta = 77.5°$

8-I
1. 1
2. 0.5
3. 0.8
4. 0.8
5. 0.6
6. 0.75
7. 1
8. 1

8-J
1. 0.5
2. 2
3. 1
4. 0.125

8-K
1. 0.1736
2. 0.5
3. 0.5
4. 0.9325
5. 0.7071
6. 0.7071
7. 1
8. 2.7475
9. 0.8660
10. 0.8660
11. 0.3249
12. 0.8480

8-L
1. 33°
2. 44°
3. 45°
4. 45°
5. 2°
6. 40°
7. 45°
8. 50°

8-M
1. 0
2. 0.3420
3. 0.5
4. 0.7071
5. 0.8660
6. 0.9397
7. 0.9848
8. 1

8-N
1. 1
2. 0.9397
3. 0.8660
4. 0.707
5. 0.5
6. 0.3420
7. 0.1736
8. 0

8-O
1. 0
2. 0.1051
3. 0.5774
4. 1
5. 1.7321
6. 3.0777
7. 9.5144
8. 28.64

8-P
1. 14°
2. 26.6°
3. 36.9°
4. 45°
5. 51.3°
6. 56.3°
7. 63.4°
8. 84.3°

8-Q
1. 20°
2. 20°
3. 60°
4. 30°
5. 30°
6. 30°
7. 45°
8. 30°

8-R
1. $\sin\theta = +0.3420$
$\cos\theta = -0.9397$
$\tan\theta = -0.3640$
2. $\sin\theta = -0.3420$
$\cos\theta = -0.9397$
$\tan\theta = +0.3640$
3. $\sin\theta = -0.5000$
$\cos\theta = -0.8660$
$\tan\theta = +0.5774$
4. $\sin\theta = -0.8660$
$\cos\theta = +0.5000$
$\tan\theta = -1.7321$

8-S
1. $\frac{\pi}{4}$ rad
2. $\frac{\pi}{2}$ rad
3. π rad
4. $\frac{3}{2\pi}$ rad
5. 2π rad
6. 4π rad

8-T
1. 90°
2. 100°
3. 180°
4. 200°
5. 270°
6. 300°
7. 330°
8. 360°